Developing Industrial and Mining Heritage Sites

Developing Industrial and Mining Heritage Sites offers a multifaceted examination of the challenges and opportunities in the development of industrial and mining heritage sites.

It does so through the case study of Lavrion, Turkey, by examining the historical process of this former mining site, which has been turned into a site of industrial and cultural heritage. It explores the ruins and monuments, which could be considered the industrial and mining heritage of Lavrion, and the restoration and rehabilitation of the mine buildings and describes the activities that succeeded through the management of a university and the finance provided for an abandoned mining site. The book also highlights the challenges experienced from the restoration of an abandoned mine site to an industrial and mining heritage site.

This book will be of interest to researchers in the fields of industrial heritage and tourism, urban conservation, mining history, post-mining, archaeology, archaeometallurgy, geology, and mineralogy, as well as policy makers and industry professionals.

Taşkın Deniz Yıldız has been working as an academician at the Department of Mining Engineering of Adana Alparslan Türkeş Science and Technology University since January 2014. At the same time, he worked at the Department of Mining Engineering of Istanbul Technical University for about four years. He completed his master's degree in 2012 and his doctorate in 2020 in the mining engineering program at Istanbul Technical University. Also in 2016, he successfully graduated from the Faculty of Law of Beykent University by studying with a 50% scholarship. *"He has 29 published articles, 18 in the Web of Science (WOS) category and 11 in other indexes. In addition, he has two published international books, one book chapter, 21 international conference papers, and 46 articles in mining sector magazines."* His fields of study are mining legislation, mining exploration and operation projects, mineral economics, mining and energy/infrastructure investment conflicts, land use in mining, evaluation of abandoned mining areas in terms of environment and urbanism, mine waste management/rehabilitation, and occupational health and safety in mines.

Routledge Insights in Tourism Series
Series Editor: Anukrati Sharma
Head & Associate Professor of the Department of Commerce and Management at the University of Kota, India

This series provides a forum for cutting edge insights into the latest developments in tourism research. It offers high quality monographs and edited collections that develop tourism analysis at both theoretical and empirical levels.

Tourism, Knowledge and Learning
Conceptual Development and Case Studies
Edited by Eva Maria Jernsand, Maria Persson and Erik Lundberg

Potentials, Challenges and Prospects of Halal Tourism Development in Ethiopia
Mohammed Jemal Ahmed and Atilla Akbaba

Diasporic Mobilities on Vacation
Tourism of European-Moroccans at Home
Lauren B Wagner

Overtourism and Cruise Tourism in Emerging Destinations on the Arabian Peninsula
Manuela Gutberlet

Pseudo-Authenticity and Tourism
Preservation, Miniaturization, and Replication
Jesse Owen Hearns-Branaman and Lihua Chen

Developing Industrial and Mining Heritage Sites
Lavrion Technological and Cultural Park, Greece
Taşkın Deniz Yıldız

For more information about this series, please visit: www.routledge.com/Routledge-Insights-in-Tourism-Series/book-series/RITS

Developing Industrial and Mining Heritage Sites
Lavrion Technological and Cultural Park, Greece

Taşkın Deniz Yıldız*

* Dr, Adana Alparslan Türkeş Science and Technology University, Department of Mining Engineering

Routledge
Taylor & Francis Group

LONDON AND NEW YORK

First published 2024
by Routledge
4 Park Square, Milton Park, Abingdon, Oxon OX14 4RN

and by Routledge
605 Third Avenue, New York, NY 10158

Routledge is an imprint of the Taylor & Francis Group, an informa business

© 2024 Taşkın Deniz Yıldız

The right of Taşkın Deniz Yıldız to be identified as author of
this work has been asserted in accordance with sections 77
and 78 of the Copyright, Designs and Patents Act 1988.

British Library Cataloguing-in-Publication Data
A catalogue record for this book is available from the British Library

ISBN: 978-1-032-52800-7 (hbk)
ISBN: 978-1-032-52801-4 (pbk)
ISBN: 978-1-003-40843-7 (ebk)

DOI: 10.4324/9781003408437

Typeset in Times New Roman
by Apex CoVantage, LLC

Contents

Foreword   *viii*
List of Abbreviations   *x*
List of Tables   *xi*
List of Figures   *xii*

1   Introduction   1

    1.1. Scope of the Study   5

**2   The Industrial Heritage Concept and Its Reuse
and Preservation**   7

    2.1. The Birth and the Concept of Industrial Heritage   7
    2.2. Preservation and Reuse of Industrial Heritage   9
    2.3. Mining Heritage and Geoheritage   13
        2.3.1. Mining Heritage   13
        2.3.2. Geoheritage   14
    2.4. Industrial and Mining Heritage Tourism   17
        2.4.1. Industrial Heritage Tourism   17
        2.4.2. Mining Heritage Tourism   19
        2.4.3. Mining and Mineral Museums   23

3   History of Lavrion & Mining   27

    3.1. A Brief Overview of the History of Greece   27
    3.2. The History of Lavrion Mining   29
    *3.3. The Effects of the Lavrion Mines on Employment and
Socio-Economics   33*

**4   LTCP Projects: Project Activities, Financing, and
      Rehabilitation and Restoration of the Site   37**

  *4.1. Industrial Crisis, Birth of LTCP,
        and Reuse Suggestions   37*
  *4.2. Management and Financing
        of LTCP Project   38*
  *4.3. Legislation on the Preservation of the
        Region and Its Inclusion in the Tentative
        List of UNESCO World Heritage Sites   41*
        *4.3.1. Legislation on the Preservation
                of the Region   41*
        *4.3.2. Immovables of Greece on the Tentative
                List of UNESCO World Heritage Sites   42*
  4.4. LTCP Projects and Aims and Scopes   43
        4.4.1. Location of LTCP   43
        4.4.2. Aims of LTCP   43
        4.4.3. Features and Scope of the LTCP Project   44
  *4.5. Conducting Projects for the Process
        of Reuse   46*
  *4.6. Reuse and Restoration of Abandoned Mining
        Plants   51*
        4.6.1. Landscape of Lavrion and LTCP   51
        *4.6.2. Reuse and Restoration Projects for the
                Buildings in Lavrion   53*
        *4.6.3. Restoration Plan and Master Plan of
                LTCP   55*
        *4.6.4. Restoration and Reuse of Flotation
                Plant   58*
        4.6.5. Restored Buildings in Lavrion   60
        *4.6.6. Lavrion Port and Industrial Activities in
                Lavrion   62*
  *4.7. Rehabilitation and Environmental Projects and
        Activities in Lavrion   63*
        4.7.1. Rehabilitation Projects in Lavrion   63
        *4.7.2. Features of Underground Waste Repository in
                LTCP   67*
  4.8. Education and Other Activities in Lavrion   69

5 Industrial and Mining Heritage and Geoheritage Characteristics of the Mine Site in Lavrion and Around It 74

5.1. *Sounion National Park* 76
5.2. *The Accommodation of Mine Workers in Lavrion* 77
5.3. *Old Mine Shafts, Adits, and Railways* 77
5.4. *Features of Mineral Deposits in Lavrion* 79
5.5. *Mineral Processing/Ruins of Metallurgy Plants in Lavrion* 82
5.6. *Geopark Project in Lavrion and Its Preservation* 85
5.7. *Lavrion Minerology and Archaeology Museum* 87

6 Conclusion and Suggestions 90

References 93
Biography 119
Index 126

Foreword

Examining an industrial and mining heritage example in the world in a detailed way can be very beneficial for those who will conduct a study on this matter in any country. Within this scope, the industrial and mining heritage area in the town of Lavrion in Greece can be examined to set an example to the world. In the literature, there have been many studies examining the historical process of Lavrion becoming an industrial and cultural site in different disciplines. However, no study at the academic level as a whole has been conducted so far. Considering this situation, to set Lavrion's industrial and mining heritage as an example to the world, from a broad perspective, the most comprehensive study among those published so far has been conducted, in which all factors affecting Lavrion becoming an industrial and mining heritage area were analyzed. Additionally, the literature on the matter of industrial and mining heritage was included to constitute a basis for the studies in the world.

In this study, the qualities of the industrial and mining heritage of Lavrion related to the mining, processing/smelting of mineral resources, and projects conducted within becoming an industrial and mining heritage are explained in detail. In 1992, under the National Technical University of Athens (NTUA), the Lavrion Technical and Cultural Park (LTCP) was founded with the contributions of the Greek state and public. LTCP has enabled the abandoned mining area and facilities to become mining heritage areas. At the same time, scientific, training, cultural, and tourism activities at this park both contribute to the region economically and enable the development of science and the growth of the popularity of mining/industry with the public.

This study describes the characteristics of the Lavrion mine site (which has become an industrial and mining heritage area), such as the mining history of thousands of years, the geology of the mineral deposits, and the mineral deposits and production. In addition, the study explains the ruins and monuments, which could be considered the industrial and mining heritage of Lavrion, and the restoration and rehabilitation of the mine buildings and describes the activities that succeeded through the management of a university and through the provided financing for an abandoned mining site. Also, in this study, the challenges experienced from the restoration of an abandoned mine site until it became an industrial and mining heritage area are described. The expenses and contributions these investment projects involved after they were completed are presented. Thanks to this study, it is possible to show in detail what great benefits the industrial and mining heritage can offer. In the study, its supporting aspects in education and training, in preserving history and culture, and for scientific development are conveyed. In this way, it was aimed to plan for suitable industrial and mining areas that can be turned into industrial heritage sites and to raise awareness among policymakers, ministries of culture and tourism, and local governments on the matter of the great potential advantages of these industrial heritage areas in the future.

The Lavrion industrial heritage area raises the awareness of historic and urban conservation among Greek citizens. There are a great number of mining/mineral processing/metallurgical plants that are abandoned and can be considered industrial heritage sites worldwide. Before these places disappear in time, the sites they are on could serve recreational, educational, and touristic purposes by being restored. Above all, there is a need for such protected industrial and mining heritage sites for the development of mining and for people to appreciate mining today. Areas that are suitable for industrial and mining heritage can be determined by considering the mineral production and processing/metallurgical plants/sites/ruins from ancient times to the present in different countries. Studies to be conducted regarding mining heritage areas will simultaneously be able to develop the mining history/industry and geoscience, along with engineering, archaeology, and many other disciplines.

This study has been prepared to be a guide for the industrial and mining heritage projects to be realized in the future in the world by giving a real-life example from Lavrion. Before this study, a book was published in Turkish to set an example for Turkey (Yıldız, 2022). This study was produced by revising and improving that book. I hope this study will contribute to the transformation of abandoned mining/industrial sites in the world into industrial and mining heritage sites and thus to the development of mining and industries, especially in developing and underdeveloped countries.

Dr. Taşkın Deniz Yıldız
July 2023

Abbreviations

AD	In the common era
BC	Before common era
CITES	The Convention on International Trade in Endangered Species of Wild Fauna and Flora
CUMP-NTUA	The Company for the Utilization and Management of the Property of the NTUA
DOCOMOMO	The Documentation and Conservation of Buildings, Sites, Neighbourhoods of the Modern Movement
EECL	Environmental Education Centre of Lavreotiki
E-FAITH	European Federation of Associations of Industrial and Technical Heritage
ERIH	European Route of Industrial Heritage
EU	European Union
Euro	€
FMC	French Mining Company (Compagnie Francaise des Mines du Laurium)
HIEM	Handicraft and Industrial Education Museum
ICOMOS	International Council on Monuments and Sites
IGME	The Greek Institute of Geology and Mineral Exploration
LTCP	Lavrion Technological and Cultural Park
LMMM	Lavrion Mining and Metallurgical Museum
NTUA	National Technical University of Athens
NTUA AMDC	NTUA Asset Management and Development Corporation
TICCIH	The International Committee for the Conservation of Industrial Heritage
UN	United Nations

Tables

4.1 Greek Real Estates in the Tentative List of UNESCO
World Heritage Sites.   42

Figures

3.1   Coin made from silver in Lavrion.   30
3.2   Metallurgical plants and railway in Lavrion.   31
3.3   Machine shops in metallurgical plant.   31
3.4   Metallurgical plants and port in Lavrion.   32
3.5   Kamariza railway tunnel.   32
3.6   Activities in mining-metallurgy plants in Lavrion.   33
3.7   Lavrion town and indigenous people.   35
4.1   Location of Lavrion in Greece.   43
4.2   Part of the restored LTCP area.   46
4.3   Post restoration LTCP area A buildings.   47
4.4   Buildings located in the LTCP region today.   48
4.5   Satellite and wireless telecommunications.   50
4.6   Laboratory of environmental measurements in LTCP.   50
4.7   Research projects at LTCP.   51
4.8   Electric power plant.   52
4.9   Industrial equipment.   53
4.10 a) Machine shop, and b) machine buildings with LMMM.   53
4.11 FMC's converted and restored buildings.   54
4.12 FMC buildings and plants.   55
4.13 a) (In FMC archives) general layout of Lavrion plants,
      1876, b) post-restoration outlook of LTCP in 2009.   56
4.14 Drone image of LTCP field today.   56
4.15 A section of houses on the front in Lavrion.   57
4.16 Three-dimensional but perspectiveless view of flotation
      building (NTUA research program).   58
4.17 Photos of the flotation building after restoration.   59
4.18 Buildings in Lavrion.   60
4.19 Church of Agia Paraskevi.   61
4.20 St. Barbara's church.   61
4.21 Buildings in Lavrion, some of which have been restored.   62
4.22 a) Lavrion boat docking, and b) Lavrion port.   63
4.23 Economic activities in Lavrion buildings.   63

4.24 Covering of waste with clay layers and HDPE liners.   65
4.25 a) Construction works for underground storage in LTCP,
       and b) underground storage adit in LTCP.   67
4.26 Informing the public in the underground hazardous waste
       storage at LTCP.   68
4.27 Photos of LTCP storage ready for hazardous waste storage.   69
4.28 Educational activities in LTCP buildings.   70
4.29 Filming in Lavrion.   73
4.30 Movie scenes shot in Lavrion town.   73
5.1   Locations and morphology of Lavrion.   75
5.2   Kamariza Railway.   79
5.3   Transport activities on the railroad.   79

1 Introduction

Lavrion is a town located 60 km south of Athens and 20 km north of Sounion in Greece that is a significant tourist and archaeological site. There is evidence showing that Ancient Greeks started mining in 3000 BC in the region where the first adits were opened in Lavrion, 50 km southeast of Athens. In ancient times, the Lavrion region became the largest mining center. The use of silver and lead ores systematically and intensively[1] started with the foundation of the Republic of Athens in BC 500. It was possible to build thanks to the immortal monuments of the Golden Age of Pericles, the tax contributions of the allies of Athens, and the silver of Lavrion. The mining and metallurgical activities of the region, which had their peak in the classical age, started to decrease between the 3rd and 6th centuries AD and finally stopped. From this date, mining and metallurgical activities could not be conducted in Lavreotiki until the 19th century. Rediscovery and extraction of the tailings of ancient silver and lead ores in the 1860s started a new prime period (Morin and Delpech, 2018; Periferakis et al., 2019; Tzeferis and Bitzios, 2020; URL-7; URL-22; URL-37). In the following years, Lavrion became the biggest metallurgical center in Greece and one of the biggest in the Balkans. The most important development in the developing industrialization of the region was the foundation of FMC in 1875. The mining company was founded by J.B. Serpieri at the northern entrance of the town of Lavrion, by the sea (Touliatos and Efesiou, 2010). Thus, the renaissance of modern Lavrion in Greece started thanks to the re-melting of the old tailing stacks in the last quarter of the 19th century, the re-opening of adits, and the development of new metallurgical activities. A period full of political, economic, and social events started, with two big companies extracting and processing minerals (URL-7). When Serpieri founded the mining company in 1864, Lavrion was an insignificant settlement. However, Lavrion turned into a town that had a population of more than 10,000 within a year. For the company employees and the local people, housing, public buildings, churches, and schools were constructed (Dermatis, 1994). Pharmacies and local hospitals operating under the auspices of the company were put at the disposal of employees. The company started using technology that was innovative for its time, such as electricity and telephones,

DOI: 10.4324/9781003408437-1

when it was bought by Syggros. Between 1882 and 1885, the company got the railway track connecting Athens to Lavrion built by bearing the cost (Katerinopoulos, 2010; URL-22). In this way, the lead and silver mine sites of Lavrion that were produced in the classical ancient age started producing again in the late 1890s after halting for centuries. Lavrion later became one of the most significant industrial centers in Greece, although it suffered severe environmental degradation due to intensive mining and metallurgical activity. However, because of the economic crisis in the late 1980s and the exhaustion of most deposits, the Lavrion mine site was abandoned. Starting from this date, environmental and economic problems started to surface (URL-14). Mine-metallurgical plants halted their activities permanently in 1989.

Shortly after its closing down, NTUA started an assertive project to turn the abandoned metallurgical area into a technological and cultural park and mineralogy museum (URL-22). In 1992, the Greek government decided to buy the whole area of the abandoned mine-metallurgical plants and transfer the management of LTCP to NTUA to create a new center of attention in high technology and culture through the foundation of LTCP (URL-11). LTCP is a pilot project of NTUA that reuses the old buildings of FMC to create a development center in the region, and to bring research and business activities together (URL-3). LTCP, additionally, is a significant project that aims to renovate an old industrial and mine region and develop new technological activities in the region, financed by the EU and the Greek state (Protonotarios et al., 2002). LTCP was transformed into a technological and cultural park through NTUA's work when it had been an industrial complex operated by FMC for more than a century (1875–1981) (Ostręga et al., 2020). LTCP aims to establish a connection between the needs and interests of the enterprises with the scientific and technological research conducted at NTUA. The LTCP area is a unique monument of industrial archaeology and architecture. It consists of 41 buildings with a total area of 25,000 m², designated as an archaeological area by the Greek Ministry of Culture and Sports (Janikian and Theodossiou, 2009).

Between 1994 and 1997, a rehabilitation project was designed and implemented by NTUA that transformed the former mine site into a scientific research, education, business, and cultural institution, LTCP. Within the scope of the Competitiveness Program financed by the Ministry of Development and EU, the NTUA AMDC undertook the project Soil Rehabilitation and Supplemental Infrastructure at LTCP to conduct a vital environmental response. Today, LTCP's services in its renovated plants support research, education, and technology. LTCP is the only specialized technological park, which is the key to modern applied technologies such as information, electronics, telecommunications, and robotics, and fairly demonstrates the former leading position of Lavrion (URL-31).

Thanks to these projects, some of the metallurgy plant buildings are still preserved today and continue presenting the phases of the production process that lasted until 1988 (URL-22). In almost all the Lavrion mine sites, many monuments/ruins belonging to the Classical era witnessed the mining activities. Entrances of underground adits and wide mine shafts to extract minerals from the ground and antiques and ruins made of some ancient mineral processing units and metallurgical furnaces, preserved splendidly, still can be seen in many places (Voudouris et al., 2021; URL-5; URL-6; URL-7). The Lavrion mines were crucial in shaping the history and socioeconomic framework of ancient Greece and the modern Greek state. It should not be forgotten that the mineralogical richness of the region is indisputable. Ore deposits in the region have always been the focal point of studies. Many mineral samples obtained from the minerals and slags in the region are unique or were discovered there for the first time. Lavrion has been an important region for both its geological uniqueness and significance and for the local and national economy. Lavrion was a catalyst for significant Greek historical events. The Lavrion region meets all scientific, educational, cultural, and touristic criteria necessaryto qualify as a geopark. In addition, Lavrion is a very popular tourist destination, considering its closeness to Athens (Conophagos, 1980; Morin and Photiades, 2008; Periferakis et al., 2019; Voudouris et al., 2021). The Lavrion mine site, which is on the Tentative List of UNESCO World Heritage Sites, is also viewed as a potential UNESCO Global Geopark, considering all the aforementioned features (Theodossiou-Drandaki, 1996, 2001; Velitzelos et al., 2003; Katerinopoulos, 2010; Papaspyridakou, 2013; Moraiti and Staridas, 2015; Papaspyridakou et al., 2017; Voudouris et al., 2019b).

The Lavreotiki region is renowned worldwide for the operation of mixed sulfide ore deposits (Pb-Zn-Ag) and the abundance and diversity of mineralogical samples (URL-12). This region contains many different ore deposits, including porphyritic Mo-W, skarn Fe-Cu-Bi-Te, carbonate-substituted Pb-Zn-Cu-As-Sb-Ag ± Au-Bi ores, and vein-type Pb-ZnCu-As-Sb-Ag-Au Ni-Bi ores. Carbonate-substituted ores of the Kamariza region were extracted in ancient and modern times and as part of a metamorphic core complex in the Atticocycladic crystal belt. Primary ore minerals of this system are pyrite, arsenopyrite, sphalerite, galena, and chalcopyrite. Galena is the principal carrier of Ag, whose maximum enrichment reaches up to 3,000 gr/ton. In addition, the presence of high-grade gold (~100 gr/ton) known and possibly used in ancient times in the nearby vein-type Clemence deposit is brought up (Voudouris et al., 2008, 2018). Even though Lavrion mines are known to be the best of their kind in Greece, there are other mine areas in Greece in terms of socio-culture and economics. They are part of the rich geographical heritage of Greece, ranging from the emery mines on Naxos Island to the operating of Au + Ag ± (Cu, Pb) deposits in Macedonia. On the other hand, with the abundance of the ore deposits in the region, compared to other regions in Greece, the mineralogical richness of Lavrion puts it forward as a significant natural landscape (Periferakis et al., 2019).

Thanks to the project activities conducted within the body of the LTCP in cooperation with various universities and research institutions from many parts of the world, the level of scientific knowledge in geology, mining, and mineralization processes, as well as environmental and pollution (containing the toxic elements lead, cadmium, and asbestos) processes has greatly increased. Lavrion has attracted a great number of researchers working on archaeology and archaeometallurgy as well as researchers on geology and mineralogy. In addition to this, Lavrion has become a unique area for training and research in various disciplines, including geological research, mineral exploration, mineralogy, geochemistry, mining, mineral processing, metallurgical processes, economy, politics, and social sciences. Most importantly, what makes Lavrion even more impressive is the combination of all the mentioned disciplines, thus providing a better understanding of the overall raw material production chain (Voudouris et al., 2021).

Industrial heritage represents the physical ruins remaining from the technological and industrial history, memories, and values. The industrial heritage of a region is also a value of the cultural heritage of that region (Mapesa, 2018). Lavrion has a significant industrial and cultural heritage. As a matter of fact, the Lavrion mines not only shaped Greek history and culture both in the ancient and modern ages (Prosser, 2019) but also history and culture in terms of the region and Europe. These mines are proof that history and geology are intertwined. Also, this region bears witness to how natural resources affect society, economy, and culture (Periferakis et al., 2019). All the features mentioned previously show that Lavrion's industrial and mining heritage is is worthy of examination as an example to the world.

This industrial and mining heritage increases the awareness of history and urban conservation, especially in Greek citizens; instills an appreciation of mining; and helps the development of many branches of science. There is not enough mining heritage declared within the scope of mining worldwide. However, there are lots of abandoned mining/mineral processing/metallurgy plants and sites that could be industrial and mining heritage sites worldwide. By renovating these places, the sites they are on can serve recreational, educational, and tourism purposes. Additionally, they can provide a revival of both domestic and incoming tourism. Above all, there is a need for such protected industrial and mining heritage sites for the development of mining and for the people to appreciate mining today. It can be said that developing countries and less-developed countries are quite behind in this regard.

Some studies in the literature on industrial and mining heritage are given in Chapter 2. These studies can be evaluated as preliminary information for mining heritage. The declaration and protection of industrial and mining heritage should be conducted simultaneously with the awareness of urban conservation and with their projects.

Areas suitable for industrial heritage can be determined by considering the mineral production and processing/metallurgy plants/sites/ruins in countries from ancient times to the present. In this study, although only the Lavrion industrial and mining heritage example is given, indirectly, it is aimed towards the determination of abandoned mine sites with the potential to be industrial and mining heritage sites in countries and guide the necessary activities for the rehabilitation/recreation of such an area, the processes to implement, and solution of the problems. It is aimed to raise awareness among policy-makers, ministries of culture and tourism, and local governments about the potential for high benefits to be obtained from these industrial and mining heritage areas in the future. In addition, this study is intended to be a guide for the public adoption of development through mining and industry. Studies to be conducted in the field of industrial and mining heritage around the world will simultaneously develop many branches of engineering, archeology, and many other branches of science, especially mining/industrial history and geosciences.

1.1. Scope of the Study

This study was prepared to shed light on the multifaceted examination of industrial and mining heritage areas in countries and the discovery of their benefits by giving the example of Lavrion. In addition, this study has been prepared to shed light on the evaluation of industrial and mining heritage potential and characteristics in countries. Thanks to the projects to be developed in this direction, industrial and mining heritage areas will be declared, which will be visited by local and foreign tourists and can be turned into an educational and scientific centers with scientific research projects. The decision processes undertaken jointly by the government, the university, and the people of the region in the achievement of reuse, restoration, and rehabilitation of the LTCP area to become an industrial heritage area and the functions, management, and finance of LTCP, and the different activities operating in this industrial and mining heritage area are explained in detail. The scope of the study within the different chapters is as follows.

In Chapter 2, topics such as the birth of the concept of industrial heritage in the world, the content of this concept, preservation of industrial heritage, reuse of the sites, mining heritage and geoheritage within the scope of industrial heritage, and industrial and mining heritage tourism and mining museums are covered. In the third, fourth, and fifth chapters, the Lavrion mines and LTCP are examined from different points of view. First, in Chapter 3, Greek history is explained briefly from Ancient Greece to the present, in accordance with mining history. Mining history is shown from ~3000 BC, when mining started in Lavrion, to 1989, when the mining sites and metallurgy plants were completely abandoned. Also, the effects of the Lavrion mines on employment and socio-economy are briefly given. In Chapter 4, the birth of LTCP projects with the effect of the industrial crisis; project activities conducted by LTCP; the aims of projects; their scope; the finance, reuse, and restoration of abandoned mine-metallurgy plants; the town of Lavrion's buildings; the aim of minimizing the negative impacts on the environment, landscaping of

Lavrion, and LTCP; legislation regarding the preservation of the region, such as including Lavrion on the Tentative List of UNESCO World Heritage Sites; and education and other activities in Lavrion are analyzed. In Chapter 5, the industrial and mining heritage, and geoheritage characteristics of the mine site in Lavrion and its surroundings are evaluated. Within this scope, Sounion National Park, located close to the center of Lavrion; sites in Lavrion where miners stayed; ancient mine shafts; adits and railways; features of mineral deposits in Lavrion; ruins of mineral processing/metallurgy plants; the geopark project and protection in Lavreotiki; and the Lavrion Mineralogy and Archeology Museum are examined in detail based on their importance.

Note

1   It is estimated that between the 7th and 1st centuries BC, 1,400,000 tons of lead and 3,500 tons of silver were produced.

2 The Industrial Heritage Concept and Its Reuse and Preservation

2.1. The Birth and the Concept of Industrial Heritage

Having high levels of industrialization by increasing production varies among countries, and it goes back to the 18th century (Falconer, 2013). The mining and steel industries became the determinant sectors for the economic development of European countries in the early 18th century. Fundamentally, these sectors started to be the representatives of economic development and urbanization after the industrial revolution. However, developing information technologies starting from the late 1960s and new energy resources changed the world economy. As a result, heavy industry lost its importance for industrial countries (Kaçar, 2016). Most of the industrial construction occurring during the industrial revolution became insufficient because of manufacturing technologies changing and developing, changes made in their production systems, and the changes in the attitudes and habits of societies and people (Casanelles, 2013; Wergeland, 2017). This situation led to the demolishing or functional loss of existing industrial buildings. In this way, efforts to involve industrial ruins in cultural heritage values came about with the effect of technical innovations that changed and transformed human life starting from the 1960s (Sirel and Çerkezoğlu, 2016; ICOMOS, 2016). Industrial leader countries decided to protect industrial buildings and the sites they were on as cultural assets (Kaçar, 2016). The approach to preserving the abandoned industrial buildings first emerged in England where building stocks were abundant (Saner, 2012; Berger and Pickering, 2018; Mapesa, 2018; Kılıç Yıldız, 2021). Subsequently, studies regarding the repurposing of industrial heritage sites in countries like Germany, France, the Netherlands, and Spain were considered a socio-economic gains for cities; thus, the same sites started to be considered industrial heritage areas (Karadağ and İncedere, 2020). An increase in industrialization in societies, a decrease in traditional manufacturing and production systems, and an increase in computerized and automated production techniques increased the interest people had in these industrial heritage areas (Xie, 2006). Thanks to the restoration and reuse of these industrial plants, the interest of local/regional/national authorities turned towards these

DOI: 10.4324/9781003408437-2

historic industrial sites (Ćurčić et al., 2015). Hence, as a result of the threats to the industrial buildings as of the 1950s, and the increase in the importance of industrial ruins that are the evidence of the business world and human life of that time, the necessity of preservation of these buildings was brought up, and the concept "industrial heritage" emerged (Sirel and Çerkezoğlu, 2016; Uysal, 2018). Acknowledgment of industrial heritage varies by country and society (Álvarez et al., 2010; Fragner, 2013; Eklund, 2018; Fontaine, 2018; Uysal, 2018).

Industrial heritage is an integrated approach to the evaluation, conservation, interpretation, financing, and management of the complex heritage of industrial cultures (Alfrey and Putnam, 1992). At the same time, industrial heritage is the field of knowledge that combines the study of construction, geographical environment, human environment, technological processes, working conditions, skills, social relations, and cultural expressions (Lin, 2013; Ifko, 2016). Industrial heritage was briefly defined by Edwards and Coit (1996) as buildings, areas made by people, and the views resulting from the industrial processes belonging to previous periods. When the definitions associated with the concept of industrial heritage are evaluated, industrial heritage includes (TICCIH, 2003, 2011; Negri, 2013):

*   The social, historic, architectural, technological, or scientific ruins in the urban and rural areas of the processes of industrial production in the past or present.
*   Tangible values related to architecture, urban planning, and engineering, and sentimental values such as memories of employees and societies, their organization, skills, technical information, and lifestyles.
*   Uses such as machines; mills; workshops; buildings; factories; sites for mineral processing and smelting; areas for energy production, transmission, and use; transportation infrastructure; and places used for social activities (housing, worship, education, etc.). It represents the connection between natural habitat and culture.

Industrial heritage, evaluated as an inseparable part of cultural heritage that is a mosaic of the perceived values of archaeological sites, history, buildings, monuments, art, or science (Güdü Demirbulat and Karaman, 2014), is an important tool in the sustainable development of societies (Labadi, 2018). The concept of industrial heritage consists of the specific architecture in which the activity of producing goods or services is carried out with mechanical tools and mechanisms (Oglethorpe and McDonald, 2013; Kaya and Yılmaz, 2018).

Many industrialized countries experienced remarkable economic growth from the 18th to the 20th centuries, with heavy industrialization and rapid urbanization. After the decline of the industrial age in developed countries in Europe, many old industrial centers (e.g. mining and steel production) became unprofitable after various periods. These changes had a profound impact on such regions. The total closure or significant downsizing of production areas triggered difficult processes such as deindustrialization, unemployment, and emigration. On the other hand, the industrial age left a significant number of industrial heritage sites, such as factory ruins and abandoned mines. These regions are places that have sought innovative spaces and practices to overcome economic, social, and environmental changes (Falconer, 2013; Harsft, 2015). The fact that industrial heritage sites are highly complex structures that are an important part of our social, spatial, cultural, and technological history necessitates a certain interdisciplinary approach in the research, evaluation, and management of these areas (Ifko, 2016).

2.2. Preservation and Reuse of Industrial Heritage

The UNESCO World Heritage List includes more than 1,000 cultural, natural, and mixed World Heritage Sites of outstanding universal value. Many of the sites face geological threats that may have adverse effects on the value, integrity, and accessibility of heritage assets (Tapete and Cigna, 2017; Cigna et al., 2018). In addition to geological risks, industrial heritage sites may also face risks and threats from a variety of sources, such as climatic or seismic risks, unnecessary interventions, and excessive tourist use. In the protection of these heritage sites, it may be necessary to analyze these risks and offer different solutions (Gámez, 2013). Climate change adaptation and disaster risk reduction are issues that concern both natural and cultural world heritage sites (Larsen, 2018). The provision of basic geological and other hazard data contributes to raising the preparedness of UNESCO World Heritage Sites' managers and the World Heritage Committee to deal with natural hazards (Kossoff et al., 2016; Cigna et al., 2018).

With the increasing importance given to urban renewal and conservation, the organizations leading the way in the goal of protecting the industrial heritage (Uysal, 2018) are as follows: TICCIH, ICOMOS, E-FAITH, ERIH, and DOCOMOMO (Hughes, 2013; Büyüktaşkın and Türkel, 2019). In addition, UNESCO participates in preservation studies (Uysal, 2018). TICCIH, in collaboration with UNESCO and ICOMOS, attempted to produce criteria for recording the represented areas of 20th-century heritage (Migone, 2013). In line with these developments, industrial heritage has started to attract attention in the academic field, especially in the last 30–35 years (Hewison, 1987; Alfrey and Putnam, 1992; Smith, 2001; Edensor, 2005; Bandarin and Oers, 2012; Douet, 2013; Cho, 2014; Xie, 2015; Wicke et al., 2018). For example Mieg et al. (2020) set criteria for both the management of industrial heritage sites ("Good Practice Wheel") and innovation areas (success factors for European Science and Technology Parks). In this study, it has been determined that the priorities of protection and innovation in the management of UNESCO industrial world heritage sites overlap, and the importance of good heritage management is emphasized.

The spread of the concept of industrial heritage brought along the preservation of significant industrial ruins in the cities encountering industrialization, the approach to their renewal, and thus their reuse as valuable assets (Del pozo and González, 2012; Yang, 2017) because unused and abandoned buildings are under the threat of destruction in time. In other words, if the updates are not made as a result of the changes brought about by some cultural, social, and technological developments, the buildings/structures that cannot keep up with these changes and that are below the standard face the danger of extinction (Gatelier et al., 2022). Buildings with historical-cultural value that have a place in the social memory are obliged to go through changes to keep up with the newly forming cultural identity. In this change, the role of repurposing comes forward in terms of giving due importance to bringing them back to life in terms of social values, by means of protecting cultural values. Repurposing is the most suitable method for using buildings by restoring them within the scope of industrial heritage under today's conditions. Thanks to repurposing, industrial buildings gain importance in terms of many aspects, such as architecture and economy. Buildings that have lost their original function maintain their structural qualities even if they lose their function. In this way, these buildings can be suitable for repurposing. Additionally, the reuse of industrial heritage areas can be examined from the perspective of the concept of sustainability. In this case, it should not escape attention that using existing building stock instead of reconstructing the buildings by demolishing them will overall provide economic, environmental, social, and cultural benefits. Thanks to the preservation and repurposing of existing buildings, the environment will be minimally damaged ecologically, and the building and its surroundings will become the center of attention economically (Büyükarslan ve Güney, 2013; Florentina-Cristina et al., 2014; Oevermann and Mieg, 2017b; Knight, 2018).

In addition, industrial buildings and their surroundings are of high importance for the preservation of collective memory, considering that they also contain shared memory for societies. Therefore, preserving them has significance for building collective memory (Önkol, 2012). Factories, one of the oldest existing structures in the city center, show the recent history of the city as an important element of the collective memory. Hence, the industrial heritage created by these historic factories also demonstrates the identity of the city. In the preservation of this identity and memory, re-functioning steps in since there is no possibility of continuing as a factory activity. Thanks to repurposing, waste of buildings with uncompleted lifetimes but loss of function can be prevented, and the collective memory can be preserved (Sadioğlu and Yürük, 2020).

Also, during the industrialization process, in the first place, residential areas were created in the immediate surroundings of the industrial buildings, with their economic, social, and physical dimensions for those working in the cities. For this reason, the immediate surroundings of these buildings were significant for the first users. Later, these areas, which offer different opportunities for the people of the region to benefit from, became public spaces that the people of the region can also benefit from, with areas that host various services. Therefore, industrial areas first carry a symbolic meaning and a deep remembrance value for the region they are located in. In addition, they are the visual and spatial traces of the urban memory and the collective memory of the country, with their contribution to the memory of the people of the region and the development and economy of the city (Önkol, 2012; Laraia, 2019).

Industrialization, which has affected a specific period in many ways through the history of countries, and industrial buildings, which are the reflection of this period on the physical space, can be considered two of the elements that represent and shape historical identity. In this case, industrial buildings should be evaluated as important building stocks in maintaining the continuity of urban memory (Büyükarslan and Güney, 2013; Wergeland, 2017). Industrial heritage buildings and planning the conservation of a sustainable city through which mutual benefits are provided for society are important. For this reason, suggestions for industrial rehabilitation should be a part of a general city project that requires a broad holistic approach embracing all urban policy areas and that promotes the reconciliation of industrial heritage with social progress and sustainable economic development (Karadağ and İncedere, 2020). For a sustainable urban transformation, work on regional and urban policies (Turk and Altes, 2010a, 2010b; Ozus et al., 2011; Rezafar and Turk, 2016, 2018, 2019; Türk et al., 2021) should be done and developed to attract innovative economies and maintain their synergies (Hansoy and Gülersoy, 2013). Concordantly, effective planning methods should be developed for land use in cultural and natural properties (Gülersoy, 2006; Yazıcı Gökmen and Gülersoy, 2018). These plans should be evaluated together with (Turk, 2002; Kilinc and Turk, 2021) city plans (Turk, 2002; Kilinc and Turk, 2021) and city preservation projects (Gülersoy, 2000, 2003, 2007; Gülersoy et al., 2008a, 2008b, 2008c, 2008d, 2009).

A lack of urban development and planning practices can also cause heritage protection to fail (Hofer, 2017). In developed countries, the preservation of industrial heritage in terms of the identity and sustainability of cities and their repurposing within the understanding of urban conservation and including them in contemporary city life (Özen and Sert, 2006; Koramaz and Gülersoy, 2011; Kilinc and Gulersoy, 2011; Rodwell, 2017; Karadağ and İncedere, 2020) has been accepted as an important approach in recent years due to the reasons stated in this and Section 2.1 (Yıldız, 2022). Many former industrial buildings all over the world, especially in Europe, have been transformed into museums, assembly and education centers, workplaces and shopping centers, social activity structures, and even residences for different purposes within the framework of the industrial heritage approach (Alpan, 2012). Transforming industrial buildings that contributed to the economic progress of countries into the reevaluating stage in the direction of the city and public interest use as the marks of industrial history is supported by the institutions and organizations stated previously (Kariptaş et al., 2015). Efforts should also be made to transform historic industrial sites into UNESCO World Heritage Sites, if possible, by concentrating on the conservation and reinterpretation of how they are presented to the public (Martin, 2013; Stuart, 2013; Kılıç Yıldız, 2021). To achieve the ambitious goals and objectives of the UN 2030 Agenda for Sustainable Development, special attention must be paid to strategies for local communities (García-Sánchez et al., 2021). Community involvement is becoming an important part of heritage management processes (van Knippenberg et al., 2022). The adoption of a policy in 2015 aimed at integrating a sustainable development perspective into the World Heritage Convention processes was an important step for heritage conservation areas. In summary, the overall message of this policy is that heritage protection is not possible without considering the local communities living in areas where industrial heritage sites are located (Logan and Larsen, 2018). Thus, the focus of industrial heritage is not only to preserve materials, tools, and artifacts but also to preserve family values and the environment and to strengthen society (Preite, 2013; Logan, 2018).

There are many challenges surrounding the preservation of industrial artifacts (Casanelles and Douet, 2013; Oevermann and Mieg, 2017a). Organizations in industrial heritage conservation are often faced with the problem of restoring very large machines or huge industrial structures. Industrial technical objects have features that make them different from other human-made artifacts (Tempel, 2013). Such technical objects and documents can shed light on the heritage potential of the mining industries and their associated processes in the broadest sense (Thompson, 2014). Recording and documenting industrial sites helps protect these areas (Albrecht, 2013). One of the most important sources for understanding and interpreting industrial history and industrial heritage sites is their building images and archives (Bergeron, 2013). For the preservation of industrial records, the records produced by industrial activities are necessary to understand not only their history and technological changes but also the history of the economic, technological, social, political, and even cultural development of the region. In this respect, industrial archives should be the cornerstone and historical/academic basis of industrial site reuse projects. In terms of industrial heritage, written information in archives, machine and tool catalogs, periodicals, and photographs of that period greatly assist in the execution of comprehensive projects for restoration and reuse (Trinder, 2013). Documenting a production line can also be useful in evaluating historic industrial heritage (Albrecht, 2013). Today, compared to the past, many new techniques and technologies can help record industrial heritage (Cossons, 2013).

2.3. Mining Heritage and Geoheritage

2.3.1. Mining Heritage

Historic industrial sites are considered valuable around the world because they hold clear clues about the processes that have developed today's technology and society. Historic mine sites are those with a special quality within industrial areas. This is because historic mine sites are special samples of industry areas, as they include not only architecture but also landscaping features related to topography and geography (Conesa et al., 2008; Reeves and McConville, 2011; Morin, 2013; Liesch, 2014; Wheeler, 2014). Therefore, open pits are also included within the scope of industrial heritage (TMMOB, 2006; Elhan, 2009). On the other hand, as explained in Subsection 2.1, the concept of industrial heritage refers to the industrial revolution and imposes a specific restriction regionally and historically, whereas the concept of production heritage includes all production periods, tools, and the ruins related to these without these limitations (Saner, 2012). Hence, the term "mine/mining (production) heritage" can be used for abandoned mine sites as well as the term "industrial heritage." In addition, considering that mining activities are generally conducted in the same site with metallurgical activities or that these activities are the continuation of each other within the scope of the mining industry, it would not be wrong to use the term "mining industry heritage" for mine sites.

A mine site can only gain value as an industrial heritage site after the end of operating activities at the site (Sutherland, 2015). A mining area can gain value as a heritage site only after the end of operating activities in the area. The types of land use that can shed light on the mining history, as well as the mining operation facilities, the buildings used, the machinery and tools in them, the geological structures associated with the mineral formation, and the production remains, constitute this heritage (Tempel, 2013; Abad, 2017; Yıldız, 2022). In fact, many mining operating methods and techniques (Agricola, 1950; Pinto, 2014; Ergünalp, 2020) applied before our age were effective in the development of modern mining operating, mineral processing, and tunneling methods (Kömürlü and Kesimal, 2016; Yaşar, 2020). In addition to this tangible mining heritage, there are also intangible production techniques, their development processes, mining, traditions, and habits of mine workers (Yıldız, 2022).

Historical-industrial areas related to mining are among the subjects that have drawn increased interest in the scientific world in recent years (Dieterich-Ward, 2018). Since the early 20th century, attempts have been made to promote mining-related areas as industrial heritage sites. Since the rise of the heritage industry in the 1980s, the number and size of mining sites managed and promoted as heritage destinations have increased significantly in the West (Oakley, 2018). After their closure, many mining activity areas can be declared world heritage sites or geoparks, considering their mining and geoheritage values. This creates an opportunity to develop a geo-mining heritage policy aimed at the sustainable use of these regions (Mata-Perelló et al., 2018). These areas are also valuable as archaeological and historical resources (Moroni et al., 2015; Florentina-Cristina et al., 2015) and can be made more interesting when evaluated together with mining heritage, cultural heritage, geoheritage, and geotourism (Ateş, 2016). Mineral deposits that were active in prehistoric times should be considered in terms of archaeological science (Fidan, 2015; Swope and Gregory, 2020). The results of excavation in the determined regions will provide important achievements in the declaration of the mining industry heritage of historic mining areas as geoheritage sites or the foundation of a mining museum.

The importance of the mining heritage can be demonstrated using evidence that can be promoted to preserve this heritage (Ahmad and Jones, 2013). Of course, mining heritage sites include not only these physical remains but also the "community spirit" and the ways this makes it possible (Power, 2008). Indeed, mining has contributed to human and technological advances over several millennia. Considering that both the physical process of mining and industry have impacts on societies, mining heritage has implications for society (Roche et al., 2021). Communities derive many benefits from reusing industrial buildings (Laraia, 2019). Therefore, while mines are developed as tourist attractions, for example, attention should be paid to the establishment of social influences and relationships, especially between local communities (Jones and Flynn, 2015). England and Wales, for example, have around 5,000 abandoned metal mine sites, many of which are protected for their ecological, geological, or cultural value. There has been some disagreement among the administration, conservationists, and local people on the types of reuse of these mining heritage sites (Sinnett and Sardo, 2020). They can also be used to develop sustainable development opportunities after mines close. As some good-practice examples demonstrate, mining heritage sites can be evaluated to create new green spaces and renewable energy generation facilities, as well as new tourism opportunities (Marot and Harfst, 2021).

2.3.2. Geoheritage

The use of minerals, which are part of the geo-diversity in the world, as useful raw materials is almost synonymous with the history of civilization. In fact, certain periods in the history of civilization have been given names such as the Paleolithic Age, Neolithic Age, Copper Age, Bronze Age, and Iron Age, considering the evolution of humanity and the great contributions of mines in these periods. Considering their role in the history of civilization, their extraction from the earth's crust, and the history of the methods of obtaining the metals they contain, these minerals are considered a part of mining heritage all over the world. The mining areas where they are located are also considered important as geosites. Geological and mining elements in a mining area are so intertwined that they are often indistinguishable from each other[1] (Manteca Martinez, 2013). On the other hand, mining heritage and geosites/geoheritage sites relate to archaeological, historical, architectural, industrial, technological, and cultural heritage (Moroni et al., 2015; Hroncek and Rybar, 2016).

Since the last quarter of the 20th century, tourists/travelers have not only been content with seeing foreign countries, cities, cultural heritage, and natural beauties, but they have also started to make trips including specific tours. Cultural tourism, unlike religious tourism, has become increasingly important in geological tourism (geotourism) (Hose, 1995), which is carried out beyond aesthetic evaluation to gain information and understand the geo-diversity of a place. Thus, geosites and natural monuments have stepped forward by preserving geo-diversity to promote geotourism since the beginning of the 20th century. Let's take a look at the definitions of some of these concepts and the differences between them.

Natural capital is defined as the accumulation or stock of all-natural entities, including air, water, earth, geology, and all living objects on the globe (Gray, 2019). Geological diversity, the non-living component of natural capital (geo-diversity) – using "geology" here in its broadest sense – is the different materials and processes forming all or part of the globe. Geological diversity is directly associated with biological diversity, as it is the base built on the ecosystem (Santucci, 2005). Passing down geo-diversity to future generations is as important as passing down the natural/environmental/cultural values necessary to protect heritage. These definitions do not mean that materials creating geo-diversity can never be touched. Humankind utilizes the materials constituting geo-diversity as raw materials for industry, energy, and construction. These factors of geo-diversity are among the natural resources that are essential for achieving the UN Sustainable Development Goals (Gill, 2017). Geo-diversity also helps integrate nature conservation with the sustainable land method in terms of planning (Serrano, 2007; Kretschmann, 2013). Geoheritage/geological heritage is a term used to describe minerals, rocks, soils, fossils, and landforms that are of significant value, justifying their conservation and proper management. This heritage includes the geological processes that continue to produce these particularly important physical objects (Brilha, 2021). The geo-diversity factors, deemed important in terms of culture, education, science, or aesthetics, constitute geological heritage (geoheritage) (Crofts and Gordon, 2014). These carry or have the potential to be important for scientists in determining the history of the world (Habibi and Ruban, 2017; Gray, 2019). Also, their locations are identified as "geosites" (ProGeo, 2011). Objects containing geoheritage may be particularly at risk due to human actions. Therefore, specific policies are needed at both the international and national levels for the protection and conservation of geoheritage

(Sánchez-Cortez, 2019; Brilha, 2021). Processes and features shaping geodiversity are also taken under protection as well as geoheritage assets such as historical and archeological places. Geoheritage draws attention to the nature consumed by humans and supports nature protection (Kazancı et al., 2017; Gordon et al., 2021). A "natural monument" is identified as a monument that contains one or more specific natural structures with uncommon or extraordinary values with their importance in terms of their natural uniqueness, representability, aesthetic qualities, or cultural values (Gürler, 2001). The International Union for Conservation of Nature (IUCN) states that for a natural structure to be classified as a natural monument, it must contain a cultural place associated with geological or geo-morphological features or natural features affected culturally or be a natural cultural place or one with ecology (URL-33). A "geopark" is named for places that contain one or more geosites and are suitable to be used for tourist purposes (ProGeo, 2011). Geoheritage areas confirmed to be in accordance with the rules stated by UNESCO, the scientific, cultural, and educational foundation of the UN, are accepted in the World Heritage List. National geoparks developing in line with the determined principles also join the Global GeoPark Network (UNESCO, 2014).

The addition of UNESCO to the global geoparks network ensures the preservation of unique geological heritage that a geopark contains. It also creates new opportunities for the development of sustainable tourism since a geopark like this attracts the attention of tourists. Additionally, in this way, tourist services, accommodation, dining, and new roads can be built. These activities have an impact, particularly on landscaping and the local population. UNESCO guidance does not allow overutilization of the richness of area through taking advantage of a political situation or socio-economic groups changing their short-term interests (Królikowska, 2017; Gałaś et al., 2018a, 2018b). Stating the accessibility of areas for geotourism is a significant phase in tourism development. Based on spatial analyses, the conditions and limitations of making geological areas accessible can be evaluated in terms of nature protection. Determining the use limitations of different environmental resources is a significant challenge, because within the geopark areas of the future, other activities such as mining[2] on a local level, agriculture, and energy investments can be carried out as well as tourism (Gałaś et al., 2017).

A UNESCO candidate geopark should be arranged in accordance with the rules that ensure the preservation of landscaping diversity and the rules of documenting, displaying, and explaining the unique outcrops and geological processes. Raising awareness and taking proper measures are required for preserving geological heritage and cultural heritage (materially and morally). Economic activities should be steered in a way that will ensure raising the environmental consciousness of the public within the framework of sustainable development, and that will create new functions in the professional service of tourism. Geopark plans should not only demonstrate the unique values of the environment but also include the provision of necessary infrastructure and educational activities to raise awareness of the local community and tourists regarding the need to protect these extremely valuable elements of the environment (Gałaś and Gałaś, 2011; Gałaś et al., 2018a, 2018c; Sallam et al., 2018; Ferraro et al, 2020; Rodrigues et al., 2021). Sustainable mining, environmentally friendly geotourism activities, and an inclusive and holistic approach with the integration of all stakeholders are needed for the geo-heritage and geopark ecosystem (Singh et al., 2021).

Effective management of geodiversity and geoheritage requires integrative, multidisciplinary approaches in different fields. Conservation requires the participation of national or local authorities and the existence of adequate legislation. Conservation requires the participation of national or local authorities and the existence of adequate legislation. There is a need for an integrated approach to the protection of geo-resources that includes cooperation between academia, public administrations, and society (Garcia et al., 2022). Geoheritage can be evaluated within the scope of natural assets (Kazancı et al., 2017; Oygür, 2021). Scientific identification processes, which are conducted in accordance with national laws and regulations and international conventions, in regions where natural assets, including geosites, and cultural assets exist are still an ongoing dynamic process. In these processes, it is required that extensive studies be conducted in each area by the experts in the area. In many countries, although conservation awareness and legal measures on cultural assets have reached a certain level, natural assets still do not have the same opportunity (Çiftçi and Güngör, 2016).

While it is very difficult for mineral and geological heritage sites to be included in the World Heritage List of UNESCO, and it is difficult to be recognized as a geosite/geopark, these heritage sites can't become world renowned in a short time. The first things to do on this matter are allocating more time by aiming to enter the aforementioned list, starting the proper project works for preserving and reusing of existing mining/geology heritage sites without losing them (Oygür, 2021), and, in line with this, appropriating funds for these projects. In some countries, the activities required for these projects can be organized in a joint working committee of the ministry of energy and natural resources and the ministry of culture and tourism (Yıldız, 2022).

2.4. Industrial and Mining Heritage Tourism

2.4.1. Industrial Heritage Tourism

Hospers (2002) states that developing tourism activities and industries in human-made areas (such as buildings) resulting from the industrial processes of previous periods can be evaluated within the framework of industrial heritage tourism. Industrial heritage tourism also includes tourists visiting industrial centers to increase their cultural knowledge (Xie, 2015; Vargas-Sánchez, 2015). Industrial heritage tourism is a largely non-profit activity primarily driven by public-sector funding and volunteers to generate economic viability (Lane et al., 2013). On the other hand, industrial heritage tourism has, directly and indirectly, an income-generating feature on a level that stands out in every area, although it can not replace the lost industrial employment. Also, it is possible to evaluate industrial heritage tourism that has importance in terms of re-regaining the reputation of old industrial areas as a preservation activity (Kaya and Yılmaz, 2018). The preservation of industrial monuments is no longer considered a cultural luxury but a necessary expenditure that promises high economic returns (Ebert, 2013). In addition to this, there are also some benefits of industrial heritage tourism (host society) for local communities (Xie, 2006). Tourism can also discourage migration to cities by employing local communities (Vuin et al., 2016). Indeed, industrial heritage areas/buildings that benefitted from economic growth in the industrial period became a resource for creating new social identities and developing industrial heritage tourism (Ballesteros and Ramírez, 2007; Yang, 2017). In this way, industrial facilities, which are seen as ungainly and ugly, will be the basis for innovative development and a successful tourism product related to industrial heritage (Ćurčić et al., 2015).

As stated in Section 2.1, the concept of industrial heritage emerged in the 1960s in line with the efforts to preserve and provide sustainability for values as evidence of the history of humans. It started to become a subject of tourism in the 1990s (Uysal, 2018). In the 1990s, the importance of the potential of this new tourism type was noticed, and a significant opportunity came about for abandoned industrial areas to come alive again (Hospers, 2002). The first developments in the name of industrial heritage tourism took place in England, where the Industrial Revolution began, but at the same time, the decline in industrial production began before all other countries. Then, countries such as Germany, France, Spain, Italy, Austria, the Netherlands, and Belgium led in terms of industrial heritage and its touristic use (Aydın, 2013; Yılmaz, 2014; Kılıç Yıldız, 2021). The first site to be included in the World Heritage List prepared by UNESCO as an industrial asset in 1978 was the Wieliczka (Poland) Salt Mines, located in Europe (Ayhan, 2009; Saner, 2012; Fragner, 2013). Then, it is seen that the development programs of industrial heritage tourism were gradually implemented in many unused regions of Europe. The EU also encourages the development of activities for industrial heritage tourists. In this framework, a project called ERIH was financed by EU funds (Hospers, 2002; Lageard and Drew, 2015). This route starts from Ironbridge in England and ends in the Ruhr coalfield in Germany. There are 19 settlements on the route where the historical assets of the coal and steel industry of the region are located (AliAğaoğlu, 2004; Hughes, 2013; Okada, 2013; Berger and Pickering, 2018). Tourism-based economic development focused on industrial heritage is targeted in the Blaenavon Industrial Landscape in South Wales, which has achieved UNESCO World Heritage Site status. In this way, local communities were encouraged economically (Jones and Munday, 2001). Each country can apply an industrial heritage strategy depending on the differences it has experienced in the historical development process (Timothy and Boyd, 2002). Thus, big industrial cities aiming to change the image of the city, diversify their economies, and form a new economy have become significant tourism cities worldwide (Okada, 2013). The Ruhr coalfield is one of them (Kretschmann, 2013; Berger et al., 2018; Berkenbosch et al., 2022). Although industrial heritage tourism has a small share in the overall heritage sector, it is estimated that over time, it will become a sector with around €2.5 billion in direct and indirect income-generating activities (Lane et al., 2013).

2.4.2. Mining Heritage Tourism

The value of a mining area depends not only on the economic benefits derived from mining activities but also on geological, environmental, social, and cultural values (Forget and Rossi, 2021). Mining has been the foundation of the development of industrial societies around the world. When mineral reserves are depleted, abandoned mines remain. At this point, the mining site, which is valued for industrial purposes, leaves its place to mining heritage (Yıldız, 2022). This site can also become a valued area in terms of tourism aspects. In this way, a new type of tourism related to industrial heritage has developed, which includes the establishment of visitor attractions related to mines, considering the societal values associated with mining: "mining heritage tourism" (Conlin and Jolliffe, 2015). It was thought that the possibility of "reusing" old industrial and mining heritage sites could help develop an economic opportunity for the recession that affected most industrial areas after the 1960s. In this direction, the focus is on mining heritage tourism and management (Perfetto and Vargas-Sánchez, 2017). Mining history and industrial tourism related to mining are not as comprehensive as general cultural heritage tourism. Due to this situation, there has been limited interest in mining tourism as opposed to other more well-known and developed forms of heritage tourism. At this point, it has emerged that tourist attractions should be included in the wider framework of heritage tourism where the tourism potential of mining areas is high (Edwards et al., 1996). Thus, mining heritage tourism has become the center of attention of the scientific community in the last 20–30 years (Hardesty, 1990; Edwards and Coit, 1996; Rybar, 2010; Tuğcu, 2012). These developments have produced a wide variety of visitor attractions such as mining museums, the opportunity for tourists to visit working mines, and the creation of various types of mining-based visitor attractions, many of which are located at or near abandoned mining sites (Conlin and Jolliffe, 2015). As a result of the drift to more individual experiences and cultural tourism, different from mass tourism in the last quarter of the 20th century in particular (Jelen, 2018), mining tourism has drawn a lot of attention in England, France, Germany, and

Spain. In all EU countries, there are 364 abandoned mines used for mining tourism (Manteca Martinez, 2013). Especially, abandoned coal mines have started to be evaluated in terms of industrial/cultural heritage (Power, 2008; Ifko, 2016; Mathias, 2018; Bozzuto and Geroldi, 2021) and tourism (Gouthro and Palmer, 2015; Lemky and Jolliffe, 2015; Niemiec et al., 2016; Vall, 2018; David and Kideckel, 2018). Most of them have become centers of attraction (Cole, 2004; Hoa et al., 2015; Price and Rhodes II, 2022; Syafrini et al., 2022). Abandoned mine sites and their heritage can be organized to be opened to mining tourism as a part of cultural tourism and geotourism, such as mining museums, geoparks, excursion routes, and visits to underground working places (Jelen, 2018; Mata-Perelló et al., 2018). Worldwide, a great number of ancient mine sites are on UNESCO's world heritage list as samples of mining heritage (URL-35), and mining tourism is broadly carried out at these sites (Oygür, 2021). For instance, in England and Germany, many tour packages are prepared directed toward domestic tourism, and, tours are organized. In this way, interest and sensitivity of the public to the mining sector are increased (Chamber of Mining Engineers, 2011; Hughes, 2013; Okada, 2013). Through similar examples, historic mine sites can advance the development of tourism by sparking the interests of people (Rentzhog, 2007).

The development of mining and mineral cultural attractions is increasing universally from both historical and economic perspectives, as reflected by the establishment of the UNESCO Global Geoparks Network and consortia for other regional geotourism activities such as the European Geoparks Network. An abandoned coal mine on the Gunkanjima Island of Japan has been included in the UNESCO World Heritage List, attracting attention to industrial heritage tourism (Hashimoto and Telfer, 2017). Similar to this site, the official recognition as a Global Geopark of the Comarca Minera by UNESCO in 2017 has increased interest in geotourism (García-Sánchez et al., 2021). The Roşia Montană archaeological mine site (Florentina-Cristina et al., 2015) is included in the UNESCO World Heritage List (URL-35). Efforts were made to include the mine site in Sardinia in the tentative list (Perelli et al., 2015).

Geotourism at mining sites ensures the preservation of mining and geological heritage and generates extra income. It shows that successful development of mining tourism or geotourism is possible through the preservation and maintenance of geotourism attractions, establishment of basic tourism infrastructure, proper promotion, and planned implementation of the site (Singh and Ghost, 2021). The most suitable mining location should be determined for geotourism planning (Sanjoy et al., 2021).

Development and planning of tourism in mining sites requires strategic implementation of public education and geological processes, cultural and historical evolution, demonstrating the value and authenticity of mining sites, and soliciting public assistance in preserving heritage (Ghazi et al., 2021). However, the key to the success of any geotourism project depends on the strong support of local governments, as well as the commitment and enthusiasm of local communities for the project (García-Sánchez et al., 2021). For example, following the closure of gold mines in Taiwan, the transformation into a major tourism experience in the presence of mining heritage has been achieved through planning strategy, promotion, and joint decisions made by the government and local associations (Shih and Lin, 2019). For the development of industrial heritage tourism, decision-makers should attach great importance to the improvement of high-priority criteria such as social benefits, resource integration, natural landscape resources, destination policy and development, economic development, cultural links, and cultural learning (Kua-Hsin et al., 2019). The use of geo-scientific, industrial, cultural, social, economic, historical, and anthropological potential for the development of mining heritage tourism is very important for rural sustainable development (Ghazi et al., 2021). There is a relationship between heritage tourism and sustainable development (Landford, 2009; Madden and Shipley, 2012; Lin and Liu, 2018). Timcak et al. (2015) discussed the aims and effects of the relevant government policies by addressing the issue of the sustainability of geotourism to indicate the development dynamics of geotourism. Metsaots et al. (2015) evaluated the mining heritage and tourism potential for regional development in Estonia's Ida-Viru District. Syafrini et al. (2020, 2022) analyzed the transformation of Sawahlunto, Indonesia, from a coal mining city to a multicultural mining heritage tourism city. In this transformation, the cultural identity of local communities was strengthened, and the common purpose of maintaining the city functioned as a means of social cohesion and integration. Jonsen-Verbeke (1999) developed a regional development plan for a mining site in the Limburg region of Belgium, considering various issues in the process of transforming from an abandoned mining area to a region ripe for tourism. Many former industrial and mining sites have lost their industrial function and are now turning to tourism for regional revival. The process of transforming from industrial (and abandoned) land into a tourist attraction is a huge challenge for city planners and tourism marketers alike. From a social and political point of view, there is a revival of cultural identity in these areas.

Mines can be venues for experiential and educational heritage tourism. Increasing numbers of tourists are looking for new experiences where they can learn first-hand about a region's lifestyle, natural features, local history, and culture. In resource-dependent areas, formerly or currently operating mines can provide this large group of tourists with opportunities to learn about the unique culture, heritage, and settlement associated with the extraction industries (Che, 2015). Depending on the social and economic history of the country visited, tourists can experience a wide variety of attractions that feature the industrial past (Gouthro and Palmer, 2015). Unlike other heritage assets, mining sites can be considered for their potential to be transformed into visitor attractions (Thompson, 2014). In recent years, many historical heritage sites in terms of geology, mining, and aesthetics have been transformed into tourist attractions (Lewis, 2020; Tang and Liang, 2022). However, many mining heritage sites may not be the main focus of tourism in the region in any case (Lemky and Jolliffe, 2015). Fundamental conflicts may arise between heritage protection and tourism attractiveness development strategies. Problems in the planning, operation, management, and financing of sites can pose a challenge for visitors to raise the profile and attractiveness of mining heritage or geoheritage tourism (Jolliffe and Conlin, 2015). For example, the Covid-19 pandemic has reduced visits to the UNESCO Global Geopark site in Peru. This has negatively affected the tourism industry in the presence of high poverty in the region (Gałaś et al., 2022). Different kinds of problems can also arise with tourism activity when there is active mining rather than an abandoned mine. Two mining operations, PowerWorks in Victoria and Super Pit in Kalgoorlie, Western Australia, faced the challenge of managing an active mine and a tourist attraction (Frew, 2015). Similarly, Lichrou and O'Malley (2006) pointed out that there may be a conflict between the two sectors in areas where mining and tourism coexist, as in Milos Island in Greece. The existence of such situations should not mean that the value of the industrial heritage will decrease or disappear completely or that other reuses cannot occur (Yıldız, 2022). Although it has been found that the tourism and mining industries generally perform better economically when they are separated from each other, it is stated that there are opportunities for managers and planners to balance mining and tourism development (Huang et al., 2011). As these examples show, in many countries, mining is mentioned along with tourism, while in many other countries, no improvement has been made for years on a level that mining history deserves. Çetinkaya (2020) summarizes what must be done on this matter as follows:

1) First, pilot projects should be determined to form a collaboration between mining and tourism.
2) The concordance of the activity area of the open pits in terms of transportation and safety should be ensured.
3) Projects should be financed under the sponsorship of public institutions or exporters' associations.
4) A cooperation center should be established between the ministry of culture and tourism and the ministry of energy and natural resources. This center should arrange existing legislation and conduct the organization of these activities.
5) Starting from primary education, children should be given museum education. In this education, our underground treasures should be taught about.
6) Promotions should be made to local administrations, especially in big cities; the problems of the pilot regions should be conveyed; their solutions should be determined; and support should be provided to the activity areas (Çetinkaya, 2020).

2.4.3. Mining and Mineral Museums

Some of the most important buildings helping to preserve mining heritage and geoheritage re mining museums (Yıldız, 2022). The establishment of museums within industrial monuments is seen by many as an ideal way to preserve industrial heritage by bringing it to new use. In this way, museums are seen as a resource for developing sustainable reuse concepts and thus protecting endangered buildings (Schaal, 2017). Another function of mining museums is to serve as a tool that provides visual learning and education for its visitors, sometimes in a fun way (Pop et al., 2004). Interactive elements are recognized as important learning and entertainment resources that appeal to tourists. Tourists express a desire to visit especially salt heritage sites that offer traditional themes but modern exhibition designs and souvenirs (Wu et al., 2015). Indeed, the Wieliczka Salt Mine Museum in Poland was included in the UNESCO World Heritage List in 1978. Also, Blegny Coal Mine Museum in Belgium, which entered this list in 2012, is one of the major mining museums. However, Big Hole in South Africa is a candidate for the UNESCO World Heritage List (Çetinkaya, 2020).

The world's first mining ecomuseum, Ecomuseo Delle Miniere E Della Val Germanasca, which used to produce talc as a mine, is located in Turin in the Piedmont region of Italy. It has been completed since 1993, using €5 million in funding from the EU (Bilgiç, 2013). In addition to this, Berchtesgaden (Germany) Salt Mine Museum (Yaşar, 2018; URL-38); Salina Turda Salt Mine (Cluj, Romania) (Çetinkaya, 2020); Tour-Ed Coal Mine Museum in America (YMGV, 2015a); Canadian Britannia Mine Museum (YMGV, 2016; Rhatigan, 2020); Seegrotte (Vienna, Austria) Underground Lake and Historic Gypsum Mine (Yaşar, 2019); Almaden-Spain Mercury Mine (Abad, 2017); Falun Sweden Great Copper Mountain (Cassel and Pashkevich, 2011), Cornwall; West Devon-England Mining District (MacLeod, 2016; Mills and McIntosh, 2021); and Nord-Pas de Calais-France Mineral Basin (Fontaine, 2018; Oygür, 2021) are the other sample mining heritage sites or museums that come to mind first.

It is known that, for passing down mining culture and knowledge to the next generations, more than 3,000 museums that only focus on mining have been established in the world, similar to the mentioned museums. It is estimated that this number surpasses 7,000 when geoscience and natural science museums associated with mining are added to these museums. Well-known museums among these are as follows: in England: National Coal Mining Museum, Britannia Mine Museum, Wilson's Mining the Museum, Durham Mining Museum, Cape Breton Miners' Museum, Peak District Mining Museum, Keswick Mining Museum, Manchester & Lancashire Collieries Mining Museums, North of England Lead Mining Museum, Nenthead Mines Heritage Centre, The Tom Leonard Mining Museum, Yorkshire

Dales Mining Museum, Black Country Living Museum, Woodhorn Colliery Museum, Ironbridge Gorge Museum, Dudley Canal Trust, Hopewell Colliery Museum, Blue Hills Tin Streams, Reigate Caves, Cornish Mines and Engines, Clearwell Caves-Ancient Iron Mines, King Edward Mine, Geevor Tin Mine and Museum, Rosevale Mine, Radstock Museum, Great Orme Copper Mines, Wheal Martyn China Clay Museum, Sygun Copper Mine, Llywernog silver-lead Mine, Big Pit Mining Museum, Welsh National Slate Museum, Wanlockhead Mining Museum, Dolaucothi Gold Mines, Arigna Mining Experience, Prestongrange Museum, and Glengowla Mines. In Germany: Bochum German Mining Museum, Rammelsberg Mining Museum, Ehrenfriedersdorf Mining Museum, Rammelsberg Mining Museum, Upper Harz Mining Museum, Morassina Mining Museum, Ore Mountain Museum, Frohnauer Hammer, Glasebach Pit, Keswick Mining Museum, Freiberg City and Mining Museum ve Altenberg Mining Museum. In the United States: California State Mining and Mineral Museum, Platteville Mining Museum, The Western Museum of Mining & Industry, Museum of Coal Mining in western Pennsylvania, National Mining Hall of Fame & Museum, Old Coal Mine Museum, Coppertown USA Mining Museum, Walsenburg Mining Museum, North Carolina Mining Museum, The World Museum of Mining, Last Chance Mining Museum, Arizona Mining & Mineral Museum, Sterling Hill Mining Museum, Lancashire Mining Museum and Gaumer's Mineral & Mining Museum. The South African Mining Museum, Kimberley Mine Museum, Diamond Mine Museum, The Gold Mine Museum, Zimbabwe National Mining Museum, Rossland Mining Museum, China Rossland Mining Museum (Chamber of Mining Engineers, 2011), Zonguldak Mining Museum (YMGV, 2015b; URL-36), Çankırı Salt Cave Museum (Oygür, 2021), Istanbul Technical University İhsan Ketin Museum, and MTA Natural History Museum (Çetinkaya, 2020) can be given as examples in other developed countries in mining (Yıldız, 2022). If the activities and functions of these museums are examined, it can be observed that they appeal to very different masses and work disciplines. Most of them are still in settlements where mining production activities are maintained. Nonetheless, underground mining operations that completed their production activities, smelters, administrative buildings, and other buildings related to mining were renovated and put into the service of the public. In museums, mining history and culture are transferred to young generations in particular and to the masses who have remained oblivious to mining (Chamber of Mining Engineers, 2011). In addition, mining and prospecting attract serious collectors and enthusiasts. The industries formed around the mines and the settlements that grew and weakened with the fate of this mineral resource are important additional elements to the attractiveness of the sites. They are an important part of mining heritage. Mineral exploration/prospecting leaves maps, examples, descriptions, and stories that may be of interest to heritage tourists and collectors even if the mine is unexploited (Dewar and Miller, 2015).

The number of open-air museums that deal with the concept of cultural heritage tourism is increasing. Mining sites are mostly located in open areas, as well as in a valley or on the slope of a mountain. For this reason, their touristic development is suitable for open-air museums. However, it is not suitable for the traditional indoor museum concepts. Revenues earned from these museums not only contribute to the social and economic welfare of the current generation but also the protection of areas that the next generations will benefit from. In Europe, the popularity of open-air museums was evidenced by 500 million museum visitors in 2004 in the 25 countries of the EU. This number is more than the entire EU population. In addition, it was determined that 33% of these people visited open-air museums (Rentzhog, 2007). The Sovereign Hill open-air heritage museum, for example, has one of the best mining-based heritage site appeals of the gold rush era in Australia. It is located on an area of 64 acres, and ~560,000 visitors annually visit this museum (Hollick, 2015).

Museums should help people by providing material on matters with which they can evaluate their past with and explore and help the public by increasing their cultural perspective. Sustaining mining accumulation and keeping the culture alive in cities/settlements where mining is intensely conducted and handing down and introducing it to the next generations are necessary. Developed countries keep mining culture alive by building mining/mineral museums and doing serious scientific research, whereas developing and less-developed countries remain behind in this area. Building mining museums in line with the ability to fulfill the needs and interests of the masses in settlement areas where mining is conducted will also be able to provide promotion of mining (Pop et al., 2004; Chamber of Mining Engineers, 2011, 2014; YMGV, 2015a). Building these museums will also enable appreciation for mining across the country (Yıldız, 2022). Cole (2008) points to tourism planning, marketing, and branding of museums, considering visitor potential and interest, so that mining museums can survive or develop economically.

Notes

1 Although this is the case, in this study, it was preferred to examine the mining heritage and geoheritage issues under different sub-sections.
2 There is a concern about damaging the areas in the world heritage sites or nearby by carrying out mining activities. On this matter, United Nations Environment Programme (UNEP) and World Conservation Monitoring Centre (WCMC) published a report together (UNEP-WCMC, 2013). The International Council on Mining and Metals (ICMM) also notified its members with a circumstance disclosure that no activity shall be conducted in these areas even if there are precautions in the EIA report that will minimize the negative effect and warned the institutions financing these activities (ICMM, 2003). In addition, a similar concern is that the increase in demand for specific minerals, hence revitalizing the mining in abandoned mine/geopark areas, may become a problematic issue for historic mine sites as a cultural resource. Indeed, increases in the demand or in the prices of certain commodities can

create an opportunity for these long-abandoned sites to be potential profit-making mine sites again (Measham et al., 2021). Although it cannot be said that it will happen in all cases, in these kinds of situations, carrying out tens to hundreds of meters of excavations can destroy archeological records. The cooperation of experts from different fields, including museum experts, archaeologists, and mining and metallurgical experts, who can evaluate the importance of the determined area, may be required to both recover the records and benefit from the ores at the same time (Ateş, 2016).

3 History of Lavrion & Mining

3.1. A Brief Overview of the History of Greece

Greece is one of the oldest civilizations in the world and is considered the cradle of western culture as it is known (URL-2). Greek history began in the prehistoric period. Little was known about the lives of the people, as there was no written language at that time. The names of the sub-periods of this period are derived from their technological characteristics: Prehistoric Ages, Stone Age (600,000–5500 BC), and Metal Age (5500–1200 BC). The Metal Age is also divided into the Copper Age, the Bronze Age, and the Iron Age (Baykan, 2015; Ehsani and Yazıcı, 2018; Tombal Kara, 2020).

During these ages, there were two different civilizations in Greece: the Minoans and Myceneans. The Minoans were named after their legendary King Minos, who lived on the island of Crete between 3000–1400 BC. They were known as a wealthy empire that feared no one. However, the king's palace had wide open spaces yet was not very well guarded. They had a well-designed economic system, recorded on clay tablets. The Minoans worshiped fertility statues and strong female figures. Their written language has not yet been deciphered. Therefore, not much is known about the Minoan culture. The Mycenaeans, on the other hand, lived in mainland Greece from 2000 to 1200 BC (URL-1; URL-2). The period from 1600 BC to 1100 BC is the Mycenaean Greek Age (Wikipedia, 2022), led by King Agamemnon fighting against Troy, which Homer tells in his epics. The Mycenaeans were known to be a very warlike society and feared threats from neighboring countries. Their castles consisted of heavily fortified walls and were usually located on hills. The Mycenaeans learned bronze craftsmanship and other handicrafts as a result of their interaction with the Minoans. In the 15th century BC, the Minoan palaces were destroyed. This is thought to be due to natural and human causes. With the weakening of the Minoans, the Mycenaeans began to expand. Next, the Mycenaeans expanded their shipping trade and conquered the Minoan palace at Knossos (URL-1).

As a result, the Minoan and Mycenean civilizations lived in the Bronze Age (3000–1200 BC) and ruled. However, these states collapsed in the 11th

DOI: 10.4324/9781003408437-3

century BC, and a "dark age" followed right after (URL-2; URL-16). About 200 years after the destruction of the Minoan palaces, the Mycenaean palaces were also destroyed. Then people migrated east, and as a result, their language was forgotten in the region. For this reason, no one knows what happened for ~300 years, and that era is called the Dark Ages. As the Dark Ages came to an end, a new Greek script emerged. Homer wrote epic poems. The *Iliad* and *Odysseus* are the oldest and best-known Greek epics. During the Ancient Age,[1] the gap between social classes grew bigger, and social unrest increased (URL-1).

City-states established by Greek tribes that immigrated to the Balkans were where the first foundations of democracy were laid (URL-19). In these cities, as of the 800s BC, a cultural and military revival occurred (URL-1; URL-2). By the 6th century BC, the Greek language and culture were geographically influential in a much wider area than that covered by Greece's lands (URL-18). Greek colonies, despite being subject to the cities they came from in terms of trade and religion, maintained control of their own in terms of politics. The ancient Greeks divided themselves into independent small communities in their homelands and colonies. These cities, which they called *polis*, formed the main administrative units in Greek territory (Hooker, 1996). City-states like Athens and Sparta have taken their places on the world map. Although these city-states had little political power, Greek language and art influenced many regions. On the other hand, the classical (or golden) age of Greece began soon after. This period caused cultural shifts in the world before it ended with the Peloponnesian War (AD 404–431) in which Athens was defeated by the Spartans (URL-1; URL-2). Athens was the most powerful city-state and Sparta was the second. These two city-states waged a war that lasted for 27 years, and in the end, Athens was defeated (URL-2). In this war, Athens with its strong navy and Sparta with its stronger land forces couldn't get the upper hand over each other in the first stage. However, after the Peloponnesian War, Sparta became the ruler of all of Greece. In 387 BC, Sparta made a peace treaty with the Persians after realizing it was losing the war against the Persians. The renunciation of Cyprus and Ionia under the terms of the Peace of Antalcidas reversed the history of Greece being undefeated against the Persians for ~100 years (Wikipedia, 2022).

Alexander the Great was one of the most powerful kings of his time in this period (URL-2). In 334 BC, Alexander the Great headed for Asia and defeated the Persians on the shores of the River Granicus Stream in Çanakkale (Wikipedia, 2022). Alexander the Great, who conquered Anatolia, Egypt, Iran, and parts of Afghanistan and India, established the Macedonian Empire. This empire ruled as three dynasties. This is known as the Hellenistic period.[2] During this time, Greek culture merged with other ancient cultures, and a new tradition was created (URL-2). Thus, the peak of Ancient Greek civilization and its greatest advancement occurred during the period of Alexander the Great. Alexander pioneered the spread of Greek culture from Macedonia in the west to India in the east, from Fergana in the north to Egypt's deserts in the south (Smith, 2015). After the death of Alexander, the empire was divided into three parts (URL-1). The development of the ancient Greek civilization stalled in 27 BC as a result of Augustus Caesar's annexation of Greece to the Roman Empire as the province of Achaea (Waybackmachine, 2009). The mighty Roman Empire headed for the territory of Greece around 205 BC. For the next few centuries, Greece came under the rule of the Romans, the famous Byzantine Empire, and the Ottoman Turks, respectively. All these influences created a unique culture in the lands of Greece. A cultural revival in the late 18th century accelerated the Greek War of Independence of 1821–1829. Greece became independent only in 1827 when Russia, France, and England intervened in the Ottoman Empire together (URL-2).

During the First World War, Greek troops fought along with the Allied powers. Afterward, the economy of Greece, which lost the Turkish Independence War in the Turkish lands, weakened. In 1936, General Metaxas was appointed prime minister by the king and quickly established a fascist dictatorship. Metaxas, on the other hand, was against German or Italian domination and refused the offers of these countries, not allowing Italian troops to cross into Greece in 1940. Economic problems led to a civil war that lasted until 1949. With the economic help of the United States, the Greek economy started to recover after this date. In 1981, Greece was included in the European Community (now the EU) (URL-2).

3.2. The History of Lavrion Mining

Although it is not possible to determine when mining activities started in Lavrion, it is accepted that they started in the Minoan Period, around 3000 BC (Economopoulos, 1996). During the Bronze Age (2800–1100 BC), the Lavrion mines provided silver, lead, and copper to all the major cultures of the Aegean (Cyclades, Minoan, and Mycenaean) (URL-7). Pericles built immortal monuments of the Golden Age of Athens with Lavrion's silver and tax contributions from his allies (URL-3). Most probably, organized mining started to develop in the 8th century BC. Silver production, on the other hand, started one century later (Economopoulos, 1996). The use of silver and lead ores systematically and intensely started with the establishment of the Republic of Athens in 508 BC, and in a short time, it reached its peak (URL-3). In the classical period (5th and 4th centuries BC), mining by the Athenian state became significant for funding the big projects of the Golden Age of Athenian Democracy and for the construction of the navy in the wars, particularly against the Persian Empire. Since the silver coming from Lavrion supplies for the famous "Lavrion owl" that is their silver coin, Lavrion mines contributed to the strong foundation of the city-state of Athens (URL-7). Thus, located on a metal-producing land, Lavrion became a dynamic center of mining and

Figure 3.1 Coin made from silver in Lavrion (URL-22).

metallurgy activities in ancient times, especially until the end of the 1st century BC, when operating activities were halted and the area was abandoned (URL-15).

Until the Persian Wars, the production of Lavrion mines was limited. However, after the battle of Marathon, Themistocles persuaded the Athenians to expand the Athenian fleet (200 triremes) around 483 BC, with the expected revenue from a large silver vein in the Lavrion mines. In this way, Athens laid the foundation of maritime power. Towards the end of the 5th century, mine production partially declined as a result of the occupation of Decelea by Spartans. However, the mines continued to be operated, although Strabo said that in his time, the mines were produced from mining tailings (URL-5; URL-6). After the classical ancient age ended, important mining and metallurgical activities in Lavrion ended as well. Following that, these mine sites remained idle for centuries (URL-3; Marinos and Petrascheck, 1956).

After the Roman Period, although there are reports regarding the economic potential of the mine area in the early 18th and 19th centuries, all of the Lavrion mines were abandoned until the second half of the 19th century (Periferakis and Paresoglou, 2019). In 1860, A. Kordellas, a young geoscientist, noticed ore minerals in ancient slags. Then, Italian mining engineer Jean-Baptiste (J.B.) Serpieri persuaded his company to utilize ancient slags in 1863. Thereupon, their note to the Greek state encouraged a second prime period of mining. In 1864, by expanding the old adits and building new ones, J.B. Serpieri established the Italian-French company Roux-Serpieri-Fressynet, which only had permission to extract ore deposits in the beginning. This company undertook the processing of ancient slags and the extraction of silver-containing lead (galena) ores. In 1865, the company opened a fully equipped plant with 18 furnaces, small mineral processing units, and a railway

(a)   (b)

Figure 3.2 Metallurgical plants and railway in Lavrion (Dermatis, 2017; URL-22).

(a)   (b)

Figure 3.3 Machine shops in metallurgical plant (Dermatis, 2017; URL-22).

(Figure 3.2). Small-scale mineral processing plants and plants housing machine shops (Figure 3.3) formed the first metallurgical industry with Castellano furnaces. Production of silvery lead began in 1865 by processing slags and (for the period) partially low-grade ores. In that period, this production was the most important industry in Greece (Periferakis et al., 2019; URL-3; URL-6; URL-22). An underground tunnel (Figure 3.5) was built in 1869, which was used to transport minerals to the port (Figure 3.4) by the first railway line in Greece. In 1871, ancient mining tailings, known as low-grade ores, were taken into the scope of state ownership. This is how the Lavreotika Affair that shook Greece began. The problem was resolved when foreign companies gave their rights on slags and low-grade ores to the Bank of Constantinople[3] (Janikian and Theodossiou, 2009; URL-3; URL-6).

In 1873, the Greek Lavrion Metallurgical Company was founded, which aimed to exploit ancient tailings. The mining rights were granted in 1875 to a new company, Compagnie Francaise (URL-6), founded by Serpieri (with main mining centers Kamariza, Soureza, and Plaka). Thus, Serpieri founded FMC, which replaced Mines du Camariza. This company was founded in the

(a)   (b)

Figure 3.4 Metallurgical plants and port in Lavrion (Dermatis, 2017; URL-22).

Figure 3.5 Kamariza Railway tunnel (Janikian and Theodossiou, 2009).

Kyprianos region (URL-3). As a result, Lavrion became one of the most important mining-metallurgy centers in Europe (Figure 3.6). Mines were reactivated by French and Greek companies in the early 20th century, essentially for lead, manganese, and cadmium (URL-5). However, after the Second World War, mining started to decline in Lavrion and some other parts of Greece. Mines were being closed down in the 1970s, whereas the metallurgy industry finally lost its function in 1990 (Janikian and Theodossiou, 2009; URL-7).

(a)   (b)

Figure 3.6 Activities in mining-metallurgy plants in Lavrion (Dermatis, 2017; URL-22).

3.3.   The Effects of the Lavrion Mines on Employment and Socio-Economics

The mining history of Lavrion started in Thorikon, one of the oldest industrial zones in Europe. Thorikon is a unique ancient mine center worldwide. As it was in the ancient ages, it combined the economical, religious, and artistic aspects of life until today. The fact that mines exclusively belonged to the city of Athens allowed the special citizens to use privilege by the Athenian government. Until the Persian Wars, the production of Lavrion mines was limited. In 483 BC, the rich mineral deposits of Maroneia (now Kamariza) were discovered. Then, adits up to 120 meters long were opened. These are deeper deposits (Janikian and Theodossiou, 2009) located in the contact zone between marble and schist. The fortune earned from these deposits in Lavrion had an effect on the course of ancient Greece civilization and directly on the cultural heritage of Europe. Its revenue, through direct rental payments and overall financial growth through indirect taxation, ensured that the city-state of Athens was largely disproportionate to the expected financial capabilities of any city-state of its era (Periferakis et al., 2019). The revenue from silver mining in 480 BC was so remarkable that Themistocles suggested it be used to build a fleet (Janikian and Theodossiou, 2009). Thanks to this, Athens build a fleet of ~200 triremes equipped with trained rowers.[4] Therefore, the revenue from Lavrion contributed to the power of Athens significantly.[5] Likely Athens would not have achieved the status of one of the leading powers of ancient Greece if the Lavrion mines had not been exploited (Periferakis et al., 2019). In addition to military matters, the fortune earned from the mines was funded for the construction of the temples in Acropolis. Columnar temples like the Parthenon were probably the most expensive buildings of the Classical Age. The costs of the gold and ivory statue of Promachos in Athens, the mural paintings, and the marble statues of the Acropolis were also partially covered by the mine revenues (Stuttard, 2013; Periferakis and Paresoglou, 2019). During and after the Peloponnesian War, particularly with the discovery of rich silver and gold deposits in Thrace and Macedonia, the importance of Lavrion mineral deposits declined (Janikian and Theodossiou, 2009).

A portion of the adits that were possessions of the state in Ancient Greece was operated in exchange for a fixed sum and profit percentage (Katerinopoulos, 2010; URL-5; URL-6). In this period, in Greece and its oversea colonies economic developments occurred, and the living standards of the people improved significantly (URL-32). According to some economic historians, Ancient Greece was one of the most developed among the other countries' economies of its period before industrialization (URL-7; URL-22). The daily wage of a Greek worker was equal to ~12 kg of wheat on average (URL-3; URL-22). This wage was ~3 times higher than the wages of a Roman Egyptian worker, whose average was ~3.75 kg per day (Schieder, 2005). However, ancient Athens had very strict mining laws, and violators were severely punished (Katerinopoulos, 2010).

In the late 19th century, Greek and French companies fundamentally supported a time of new riches in Lavreotiki. As well as the development of the mining industry in Greece, they had an impact on the foundation of the town of Lavrion and its features. As a result of the workers' settlement in 1867, at the beginning of the 20th century, Lavrion had become a city of 10,000 inhabitants[6] (Figure 3.7). In the same year, compared to the employment numbers across the country, 1,200 employees, considered a record number, were employed in the Lavrion mines (URL-3; URL-6; URL-22). With the foundation of mine-metallurgy companies, the town of Lavrion started to expand. Lavrion was a small settlement place for employees. However, in a short time, Lavrion became the first "company town" in the late 19th and early 20th century in Greece. Indeed, Lavrion is the single sample in the vast region of Greece and the Mediterranean of a town company in an isolated area to meet the needs of the industry. The companies provided their employees with health services and built schools and churches. Additionally, they constructed port facilities and the Athens-Lavrion railway (URL-7; URL-22). Two companies in Lavrion were responsible for running the city. Houses and stores belonged to the company. People would get treatment at the hospitals and drugstores of the company, and they could obtain medicines. Since life in the town of Lavrion was closely associated with the industries of the region, naturally it developed relatedly with them. The first serious economic crisis emerged in the 1880s and 1890s when the price of lead declined (URL-3; URL-22). Although the companies had financial and administrative benefits for local communities, both foreign and Greek investors' conflict over mining rights and stagnation of mining policies have adversely affected national economic policy. Actually, this crisis in Lavrion became an important factor that led the Greek state, which declared its public bankrupt in 1893, to its

Figure 3.7 Lavrion town and indigenous people (Dermatis, 2017; URL-22).

financial collapse (Periferakis et al., 2019). The First World War idea deepened this crisis further. Afterward, in 1930, the Greek company sold its plants. Towards the end of the 1920s, the population of Lavreotiki municipality declined by 50%, whereas rums (Greeks) came from Turkey as a result of the exchange in accordance with the agreement made between Turkey and Greece in 1922, resurrecting the population of the city. In the mid-1950s, Lavrion went through a new period, which lasted for several decades, characterized by the development of new industries such as electrical power generation, textile industry, army industry, and tool/household goods manufacturing. Mining companies, on the other hand, following a declining course after the Second World War, began to cease mining activity in 1977–1981 and began to reduce their mineworkers by 1981. The company halting its metallurgy activities in 1989 coincided with the de-industrialization period increasing in the town. In the late 1980s and early 1990s, Lavrion came face to face with a new economic crisis cycle as a result of the large-scale de-industrialization all over Greece. Tens of factories halted their activities. According to K. Pogkas, who was mayor of Lavrion from 1975 to 1994, "between 1990 and 1993 70% of the industry left Lavrion. The city was therefore economically bankrupt. In 2.5 years, 2500 out of 3500 people in Lavrion left the city. Lavrion became a city in despair" (Pogkas, 1996; Chatzi Rodopoulou, 2020; URL-3; URL-22).

Notes

1 The Archaic Period is the period in Greek history from the 8th century BC (the continuation of the Greek Dark Age) to the 2nd Persian attack in 480 BC (Wikipedia, 2022).
2 The Hellenistic Period started with the invasions of Alexander the Great. Greek influence reached its peak in the ancient world in this period. This period followed the Classical Greek Period (Wikipedia, 2022).
3 The mining company illegally bought ancient slag loads from Keratea municipality. Thus, the company started to extract silver, which is not economical to recover, using old techniques. As a result, the company violated the mining license obligations given to the company by the Greek government. Greek courts issued a sanction order, which ordered the mining company to pay a high level of compensation. This situation prompted the embassies of Italy and France to interfere on behalf of the company and request to drop the legal actions from the state. However, the Greek government changed its stance due to the naval blockade imposed by French and Italian warships. In 1873, Andreas Syggros bought the company and changed its name to Lavrion Metallurgy Company. Syggros then persuaded the locals to buy his own company's capital stocks without par value. At the same time, he pressured the Greek state to lower the annual taxes on his company. On top of that, he reduced slags extraction and silver production. Meanwhile, Serpieri founded the Compagnie Française du Laurium, which succeeded in acquiring the right to exploit the region's mineral deposits (Dermatis, 1994).
4 This naval strength of the Greeks prevented the Persian advance (Periferakis et al., 2019).
5 This situation created a dangerous conflict between the city-states of Athens and Sparta, which were dominant back then. The balance of power, for this reason, could only change through war (Periferakis et al., 2019).
6 Lavrion's population was 10,007 in 1907 (URL-5).

4   LTCP Projects

Project Activities, Financing,
and Rehabilitation and
Restoration of the Site

4.1.   Industrial Crisis, Birth of LTCP, and Reuse Suggestions

In 1917, the Greek company took a break from its activities due to the stock market scandal, the strike of employees, and the gradual decrease in tailings through which mineral recoveries were conducted. The factories of the company were sold in 1930. After the Second World War, mining started to decline in Lavrion and the other parts of Greece. In the mid-1950s, a new era started for Lavrion. This new era lasted for a few decades, and new industry branches developed. The industrial crisis in the 1970s and 1980s affected the most significant centers in Greece, including Lavrion, which enabled the development of Greek industrial activities. In 1977, FMC, which had been active in the region for more than a century (1867–1989), started operating again with the effect of the crisis, even though it halted its mining activities. Seven years later, after a series of domestic disturbances and failed restructuring efforts, the mining and metallurgical activities of the company stopped completely. The town of Lavrion came to face a severe unemployment problem that caused social disintegration. In this way, Lavrion encountered a new economic crisis cycle as a result of the de-industrialization that took place everywhere in Greece. Tens of plants halted their activities, and more than 20% of the population left Lavrion due to unemployment. As a result of the de-functioning of the metallurgical industry in 1990, the settlement and destruction of FMC buildings were planned during the socio-economic crisis. This destruction plan, which was about to deprive Lavrion of its value that was intertwined with its industrial history and core identity, was prevented by a joint venture of the municipality, the local people, and NTUA. Parties had to take significant steps to protect the buildings from destruction and offer a new life: first, ensuring that the Greek state would buy these buildings and the area and transfer them to the Ministry of Culture and NTUA; second, designing a new program suitable for the buildings; and last, getting financing for the necessary works to be done to transfer and reuse the buildings. Although it was a challenging attempt, the determined efforts of the collaboration were fruitful:

DOI: 10.4324/9781003408437-4

In 1992, with the decision of the Institution Senate of NTUA (School of Architecture and the School of Mining and Metallurgical Engineering) academicians, the project of rescuing and restoring the former FMC plants with the suggestion of establishing a technological park and mineral and metallurgy museum was presented to the Greek government. For NTUA to establish a technological and cultural park, all the plants were bought by the Greek state. In the 1980s, before the factory was closed completely, a clear and long discussion took place about the possibilities of reusing the plants and equipment of the mining company for various activities. From the mid-1980s to the early 1990s, this discussion continued. Academicians in the field of mining and metallurgy engineering, municipality officials, and the company managers participated in these discussions where the historical value of the region and the factory buildings are widely agreed upon, and the necessity of preserving their historical, architectural, and technological features was emphasized (Chatzi Rodopoulou, 2020; URL-3; URL-6; URL-7; URL-22; URL-31). In these discussions, two options were determined (URL-3; URL-22):

1. First, a call was made for the restoration and reuse of the mining facilities and the area surrounding them for various social and cultural purposes such as theaters, museums, exhibition centers, rest areas, education-training, sports, and spare time activities.[1]
2. The second option was more sensitive to the historical background and process of the technology in the mine site and plants during the operating processes following each other since the mid-19th century. This option emphasized the importance of preserving the technological and historical identity of the buildings through an innovative attempt that aimed to renovate these plant buildings and reuse them as a technological and cultural park. In this way, it was thought that LTCP would revive the historical identity and collective memory of the region as a production site since ancient times (URL-3). On a more pragmatic level, LTCP was thought to contribute to the technological advancement of the manufacturing industries of Athens through technology transfer, dividing activities, creating new job environments, and developing its infrastructure and new applications/product innovations. Hence, the idea of LTCP was born (URL-3; URL-22). In this direction, LTCP was founded in the former settlement place of FMC in Lavrion in 1992 (URL-14).

4.2. Management and Financing of LTCP Project

The Greek government appointed NTUA in 1994 to restore and reuse the abandoned mining and metallurgical site buildings as a science and cultural park. After being sold to the Greek state and given to the Ministry of Culture, this area was allocated to NTUA by its nature. Within the scope of the 2nd

Support Community approved in March 1994, financing of 3.6 billion drachmas was prepared and presented to the Attica region. Twenty-five percent of the national contribution in this budget was covered by the Ministry of Development, with the personal initiative of the Greek prime minister himself. For activities in LTCP, this resource was allocated within the scope of the Competitiveness Programme financed by the Ministry of Development and the EU and the implementation of the 3rd EU Framework Programme[2] (URL-31).

Until 1996, the LTCP was governed by a special senate committee and the Company for the Utilization and Management of the Property of the NTUA (CUMP-NTUA) that formed since then. NTUA invested 60,000 € in 1996 for the establishment of the CUMP-NTUA, which would also provide administrative and technical personnel (Damigos and Kaliampakos, 2012; URL-3; URL-22). A committee was also formed for the management and promotion of the Lavreotiki Geopark (Periferakis et al., 2019; Tzeferis and Bitzios, 2020). By mid-1995, the administrative procedures for the establishment of the LTCP and the restoration of the plant buildings were completed. Relevant funding from both EU and national funding sources was determined as ~5.19 billion drachmas (€15.23 million). The program for which this fund would be used included the restoration and renovation of 17 of a total of 42 buildings constructed during the period 1876–1898, the construction of two new buildings, the reorganization of the surrounding area, and the rehabilitation of part of Lavrion's ~10 decares of heavily polluted/degraded land (URL-3; URL-14; URL-22). In the first stage, it was decided to finance the entire project by the European Community (€15.23 million). The first phase of the transformation of the buildings and the decontamination process emerged from a combined plan. After the decision to increase the investment, 25% (€5.7 million) of the investment required to finance these activities was covered by the Greek state, while 75% (€17.1 million) was provided by the EU funds. All projects were conducted by the academic community. More than 150 professors, researchers, and students participated in the research and projects (Fotiou et al., 2011; URL-3; URL-14; URL-22).

However, the financial crisis that negatively affected Greece in 2008 had a major negative impact on the functioning of the LTCP and deeply affected its progress. Additional funding was not available to complete the reuse of the rest of the buildings. Compared to 2007, the rented areas declined to 30% and working areas by half. Considering that the main source of revenue of the LTCP is rent, this development led to serious problems in the operating and maintenance of LTCP. Last, the ongoing strategic development projects of the region, which were expected to increase the positive effects for the LTCP and the town of Lavrion (such as repair activities on the railway line extending to Lavrion Port), were halted. Despite all these negatives, in this intense economic crisis, with the determined support of the local people, the efforts of the LTCP management, and the relevant personnel of NTUA, an important success was achieved once again. In 2010, a highly significant project was started, which had been delayed for almost 20 years, involving turning the Machine Building into the LMMM between the building blocks. These works with a budget of € 2.7 million were financed by the Attica region. The renovation project of this museum is a historical, technological, cultural, research, educational, and developmental project that provides a rare chance to attract a wider audience (Dermatis et al., 2010, Urban Environment Laboratory, 2009). In addition to this, it was planned to transport, preserve, and open the Lavrion Historical Archive to the public. These precious materials, which are currently stored in the Machine Building, have increased their importance further because it is the largest industrial archive in the country. Along with the completion of the aforementioned projects, the priorities were determined in the strategical planning of LTCP (2010–2015): decontamination of the building Konofagos, a significant project delayed due to financial and bureaucratic reasons; drawing attention to EU financing programs and their effective uses; strengthening cooperation with local organizations such as NTUA laboratories, professional associations, and the Lavrion Port Authority; and maintaining proven profitable activities consistent with the character of the LTCP (Chadoumelis, 2015).

NTUA built the LTCP area with local community support and in the presence of existing scientific experience. Thus, it developed a new model of regional, socio-economic, and cultural development based on technology in this field. This model reflects the unique characteristics of NTUA. As a technical university, NTUA does not observe or perceive technology in the way it is practiced by the market. A technical university generally perceives technology not as a commercially producible input to production processes but as a specific way of thinking that combines human skills, human understanding, and productive imagination. With this perception, LTCP wasn't designed to simply or only be a "profit-making" pole of regional macro-economic growth but a tool that was built socially to develop the cultural capital containing the human and intellectual keystones of the region and the new economy. In other words, the LTCP project was perceived as a means of transforming the abandoned mining company's buildings into a practical, self-sustaining (i.e. self-paying) development process model for the future. In this regard, an important difference between LTCP and other typical Greek science and technological parks is that LTCP sees the prospects of tenant companies as an organic part of an integrated socio-technological, commercial, and cultural environment. The aforementioned environment consists not only of technology transfer, interactive technological learning, business competition, and profit maximization behavior but also of social and cultural values and norms that are inextricably linked to the emerging new knowledge-based economy (URL-22; URL-31).

In Greece, the territory of Attica is intertwined with the existence of two organizations representing the region itself, the technology center (LTCP) and the mechanism for establishing the business development center (BIC). The origin of the project is based on the awareness that the regional growth in the MED region[3] is possible through developing a "knowledge-based economy." From this, the decision was made to form a partnership in which three main units are represented: economic operators, innovative operators (in some cases, these two are combined at business incubators or technological parks), and competent regional authorities (URL-9). The Attica region is a secondary self-governing unit of local government. The region is responsible for the planning and implementing of political decisions regarding the economic, social, and cultural developments at the regional level according to national and European politics. Within the scope of its responsibilities, the Attica region participates in the implementation of many European projects within the framework of various EU programs and initiatives (URL-10). Within the framework of the LTCP project, it is expected to provide consultancy services to creative industry enterprises interested in Greek and European research programs.

4.3. Legislation on the Preservation of the Region and Its Inclusion in the Tentative List of UNESCO World Heritage Sites

4.3.1. Legislation on the Preservation of the Region

The existing legislation regarding the preservation of the region is as follows (URL-12):

- In 1974, a big part of the Lavreotiki region was declared a national park (Sounion National Park) according to Greek laws. For this reason, the region is under a certain legal protection and management regime.
- Both divisions of Lavreotiki were included in the Natura 2000 Network sites (Gr3000005 ve SPAGR3000014) list.
- Law 1650/86 "For the protection of the natural environment".
- Law 2742/99 "For the sustainable development".
- Law 360/76 "For the land planning and environment".
- Law 998/79 "For the protection of forests and forestal in general extends of country.
- Habitats Directive 92/43/EE for the conservation of wild species of the flora and fauna and their habitats.
- UNESCO's Convention (Paris, November) "For the Protection of the World Cultural and Natural Heritage", ratified by Greek Law 1126/30-1-1-1981.
- Birds Directive 79/409/EE for the conservation of bird species.
- Archeological legislation and other Greek laws referring to the protection status of the landscape, history, and cultural issues.

- A Presidential Decree (Gazette No 125/D/1998) has also been issued for the broader Lavreotiki area (Urban Control Zone for the Lavreotiki Area) in which zones with specific land use and building regulations have been determined.

- The protection of Lavreotiki forested areas is determined by the Presidential Decree: "Designation of Protected Mountainous Zones of Lavreotiki Peninsula" (Gazette No 121/D), which is amended with Law 3212.

(Gazette No 308/A/31.12.2003)

4.3.2.   Immovables of Greece on the Tentative List of UNESCO World Heritage Sites

All LTCP industrial plants, including on-site equipment, were declared to be listed by the Greek Ministry of Culture in 1992. LTCP constitutes one of the largest and most important industrial heritage sites in the historic content–listed industrial monuments in southeastern Europe (Efesiou, 2011). While the administrative region of Lavreotiki was declared a national park, the Lavrion region was recognized as having outstanding universal value and was found to comply with UNESCO's criteria for inclusion in the World Heritage List (Migoń, 2018). In this way, the Lavrion mine site was added to the Tentative List in Greece in the last revision of UNESCO World Heritage Sites dated 31/01/2003 (Table 4.1). In Table 4.1, the representation is as follows: N: Natural, C: Cultural, and N/C: Natural/Cultural combined quality of the heritage site.

LTCP consists of smelters, storage areas, offices, and so on, with a total construction area of ~25,000 m². It consists of 41 stone buildings. These FMC buildings form a unique work of industrial archeology and architecture. In LTCP, all plants and a great majority of machine equipment were preserved. The buildings were mostly built from 1875 to 1940, and they hosted the industrial activities until 1988. In this way, all industrial plants (including the buildings, sheltered and open-air installments, machine equipment,

Table 4.1 Greek Real Estate in the Tentative List of UNESCO World Heritage Sites (URL-7; URL-34).

Archaeological Site of Nikopolis	31/01/2003	K
Archaeological site of Philippi	31/01/2003	K
Gorge of Samaria National Park	31/01/2003	D
Lavrio (Ancient Laurion)	31/01/2003	K
National Park of Dadia – Lefkimi – Souflion	31/01/2003	D
The Area of the Prespes Lakes: Megali and Mikri Prespa which includes Byzantine and post-Byzantine monuments	31/01/2003	D/K
The broader region of Mount Olympus	31/01/2003	D/K
The Palace of Knossos	31/01/2003	K

and related accessories) were listed as historical artifacts by the Ministry of Culture (I.1469/1950). Thanks to this, these plants were transformed into a technological and cultural Park (LTCP) under the administration of NTUA (URL-23).

4.4. LTCP Projects and Aims and Scopes

4.4.1. Location of LTCP

Lavrion is a town in the southeastern part of Attica and is the center of Lavreotiki municipality. It is situated ~60 km southeast of Athens, southeast of Keratea, and north of Cape Sounion (Figure 4.1). Lavrion is located in a bay overlooking Makronisos Island to the east (URL-8). LTCP is ~45 km away from central Athens and is on the Lavrion coastal road just outside the town of Lavrion.

4.4.2. Aims of LTCP

The LTCP project aims to establish a connection between the needs and interests of the enterprise doing scientific and technological research conducted in NTUA and the continuity of the education process with the related experience achieved there. Also, the LTCP project aims to culturally advertise the history and heritage of the plants it has, as well as the history and culture of a broader area of Lavreotiki (URL-3; URL-6). It was decided to have two important aims of turning FMC buildings into a technological and cultural park: museums, education, and culture as well as research and high-tech

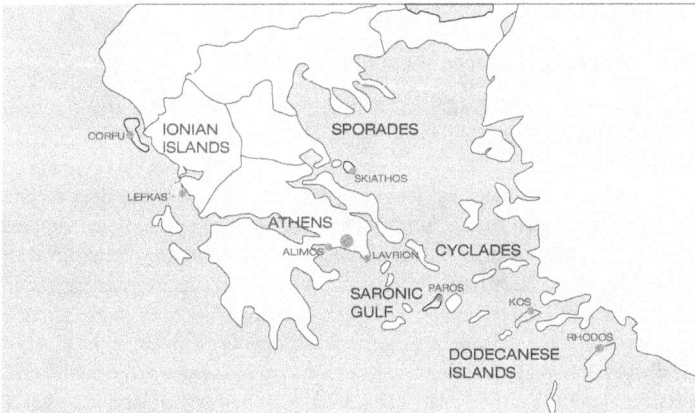

Figure 4.1 Location of Lavrion in Greece.

production on the history of buildings. Thus, thanks to these buildings, it is aims to host research teams, high-tech companies, and business incubators, as well as education, museums, archives, cultural events, and tours in the long term (Fotiou et al., 2011). The multiple goals of LTCP can be summarized as follows (URL-3):

• Contributing to creating and developing commercial activities, creating a useful workplace environment for its function, and encouraging this system in new technology fields. Evaluating LTCP sites and special factors that could support such a system with recent experiences.
• To highlight the cultural context of the place and associate it with the development demands of the region.
• To create suitable institutional and financial conditions for the transfer and commercial development of knowledge in the university (an environment of mutual support and empowerment).
• To bring the university, production, and society into developing cooperation not only at the national level but also at the international level.
• To design and implement a new communication system, as well as the existing one, that will allow enterprises to establish their operations outside the metropolitan area without compromising their needs in the plants.

LTCP is a project that consists of link mechanisms, benefits, and services to smooth the way for creating modern entrepreneurship. This project aims to provide support and continuity in technological research and development projects. The fact that the project creator and manager is the university has maximized the potential to encourage knowledge-based entrepreneurship. This project was developed in the region that is accepted as the biggest and longest-term industrial experience in Greece (URL-3).

4.4.3.   Features and Scope of the LTCP Project

LTCP covers an area of ~25 ha. A little over a third of the area has been restored. This area functions as a high-tech business and cultural center for southeastern Attica. LTCP's services and renewed buildings continue to support research, education, and technology. Today, LTCP is the only technological park in Attica specializing in areas that are the keys to modern applied technologies such as information and electronics technology, telecommunications, robotics, laser technology, environmental technology, energy, shipbuilding, and marine technology (URL-3; URL-31). For the establishment of the latest technological enterprises, a row building is available in the plants of LTCP. Research and operating activities are enabled in two basic methods: as part of an investment project of the relevant company or under a lease agreement for a particular building to adapt the characteristics of the existing area to their specific needs. In the second case, the entire project is monitored and finally approved by NTUA's Technical Department (URL-3). The innovative work of new technology companies in operating centers such as LTCP is a powerful driving force for them to effectively transform into commercial and business success. Through its supporting infrastructure and services, LTCP significantly contributes to the fastest possible union and development of business/enterprise types such as (URL-24):

• The application of innovative ideas, products, services, and processes, as well as the use of scientific and technological research results in the business world.
• Ensuring efficient connection of information production organizations with enterprises.
• Development, renewal, and expansion of the production, supply, and distribution methods, as well as the range of products and services.
• Introducing new methods of business administration and management.
• Acquisition and dissemination of new knowledge, provision of all kinds of scientific, technological, educational, and consultatory services, in addition to the provision of specialized personnel training for companies or any natural or legal person interested in general.
• Attracting the attention of enterprises and business departments and establishing them or establishing relations with outside companies.
• Attracting foreign investment in high-tech sectors.

In addition, LTCP is an important innovation center for the Attica region, which is a milestone for the development of Lavrion (URL-13). It hosts innovation centers and thus newly established units within the LTCP to provide the following (URL-24):

• Creating and developing organizational diagrams to complement and secure their innovative work.
• Developing and testing the innovative use plan.
• Implementing the innovative use plan.

In cooperation with the local community, NTUA has also undertaken a holistic cultural project consisting of many parameters (URL-23).

• Recovering, preserving, and restoring a big industrial heritage area for Greece.
• To upgrade the productive, creative, and sustainable content of historic buildings and put them into reuse (Figure 4.2).
• Recovering the intangible heritage associated with the city; carrying out relevant research in collaboration with local, national, and international

Figure 4.2 Part of the restored LTCP area (URL-22).

organizations; and disseminating it in Lavrion as well as in the larger community.

- Hosting, organizing, and producing educational and cultural activities by emphasizing the culture supported by technology and innovative ideas.
- Supporting young artists, scientists, and innovative cultural activities.

4.5. Conducting Projects for the Process of Reuse

The process of reuse was shaped through a series of decisions: First, it was decided to document, evaluate, and preserve the entire LTCP area, with the entire asset of the on-site production line, as well as the historical archive of on-site historical objects and industrial equipment. In another decision taken in parallel with this, even if there was a dilemma about choosing between the areas to be operated as a museum and the areas to be used for rental, a solution was found (Chatzi Rodopoulou, 2020). In another decision, allocation of reuse containing a small part of research and technology, production, education, culture and entertainment, and accommodation was discussed. An economically autonomous project delivery that needs no external financing for operating was available within the basic project decisions. To enable this, an activity plan by the NTUA project team and external consultants was designed. During the preparation of the studies, the project stakeholders were faced with another situation that was not given much importance at the beginning: The issue of decontamination of buildings and soil, which was an unknown area in Greece at that time, was handled with scientific diligence in two stages (1995–1997 and 2003–2009) (NTUA, 1997; Fotiou et al., 2011; Kaliambakos, 2015). With the desire to produce high architectural quality and social compliance within the current budget, an in-depth analysis of the historical development of the buildings determined the strategy of reuse and spatial use. The large LTCP area was shortly divided into three regions corresponding to the three generations of the industrial plants. Due to its historical features (1876–1895) and its central location, it was decided to start the project in Region A (Figure 4.3). The reuse works of this area of 12,000 m^2 consisting of 14 buildings were started in 1995 and completed in 1999 (Touliatos and Efesiou, 2010). The reuse was accomplished with the direct participation and oversight of an interdisciplinary team predominantly by NTUA professors, researchers, the Faculty of Architecture, and the Faculty of Mining and Metallurgical Engineering (Chatzi Rodopoulou, 2020). Today, a general map of the buildings of the institutions in the LTCP region is given in Figure 4.4. On the site, there are more than 20 host institutions in three categories: NTUA laboratories, young

Figure 4.3 Post-restoration LTCP area A buildings (URL-14).

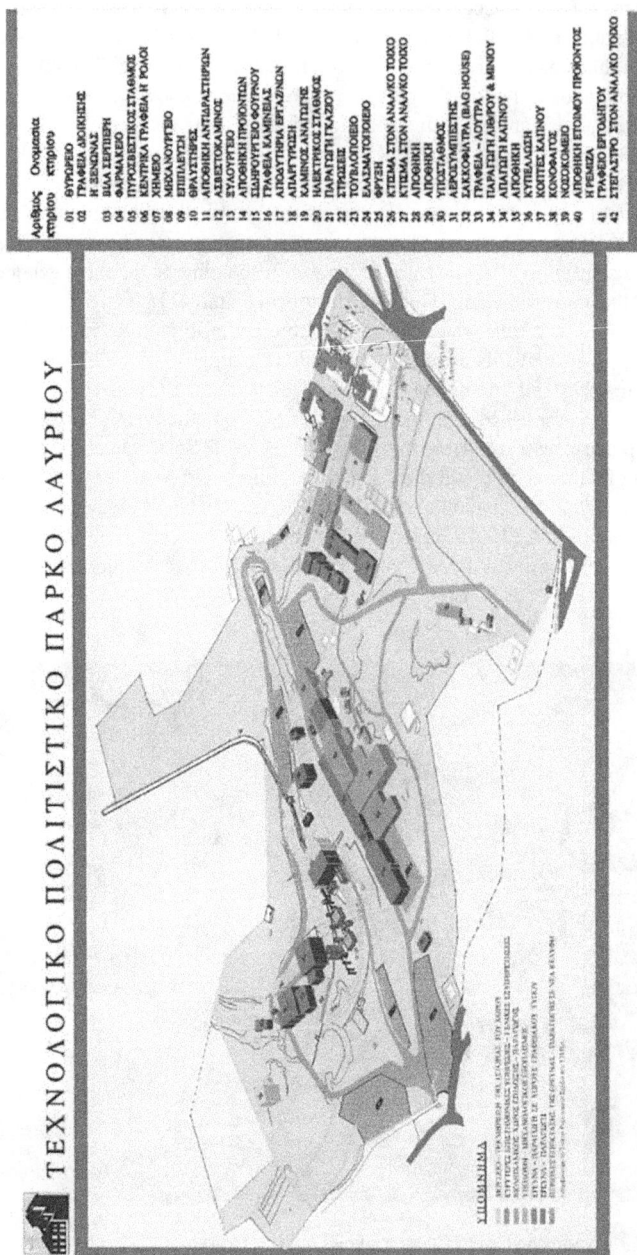

Figure 4.4 Buildings located in the LTCP region today (URL-3; URL-27).

and innovative companies, and educational institutions (URL-13; URL-27).
Institutions/companies in the LTCP region are as follows (URL-13):

* Companies involved in the production, development, and implementation of innovative ideas, products, and services in the fields of modern technology.

* NTUA laboratories that conduct the function of promoting the know-how produced in NTUA and transferring it from the university to the business world and society.

* Educational institutions such as the Environmental Education Centre of Lavreotiki and the Handicraft and Industrial Education Museum.

LTCP officially opened its doors to the public in 1999. This situation was received as a step towards the renewal and reuse of the region. From then to 2015, more than 170 cultural activities were organized, including theatres, concerts, and art activities in the plants brought into reuse. In addition, more than 20 conferences and more than 80 scientific meetings have been held here, and ~2000 students visit LTCP buildings annually. These activities are organized without external financing, with a budget of 300000 €/year, by a staff of only ten people (Kaliambakos, 2015). LTCP operates by CUMP-NTUA as an individual legal entity with special rights. The only share of the company is owned by NTUA (URL-21). The project, having innovative characteristics compared to Greek standards, drew the attention of the private sector in a short time. As of 2008, more than 70% of the available space is leased specifically to high-tech companies. Thanks to significant infrastructure improvements in the larger region of Mesogia, LTCP has put the old industrial city back on the map and presented it with new opportunities for socio-economic recovery (Kaliambakos, 2015). LTCP is a place where many technological developments take place and innovative ideas are successfully blended with high industrial aesthetics. In this area, the aim was to conduct a series of necessary and carefully analyzed activities that will provide revived research and educational and cultural activities. LTCP is a place that aims for the future and hosts research and innovative projects focused on the latest technology. Since 1998, LTCP has hosted more than 40 start-ups/companies, many of which continue to operate in industries such as satellite and wireless telecommunications (Figure 4.5), nanotechnology, and advanced materials, in addition to the residential research programs run by NTUA. Fundamentally, this versatile technological and cultural park is a center of attraction for interdisciplinary scientific and commercial activities open to researchers and entrepreneurs (URL-15). The buildings of LTCP bring a significant human mass together within the scope of science, technology, and culture. In this way, it hosts a series of organs that make interdisciplinary cooperation easier. Also, the support provided by NTUA

Figure 4.5 Satellite and wireless telecommunications (URL-24).

(a)   **(b)**

Figure 4.6 Laboratory of environmental measurements in LTCP (URL-26).

ensures absorbing the information in the most effective way and transferring it instantly with modern solutions (URL-13; URL-27).

Today, a high level of research is produced in LTCP plants, both by institutions in the field and by LTCP personnel who are actively involved in various research projects (Figure 4.6). LTCP, at the same time, is evaluated as a base research-development institution for new information products or the background of scientific-experimental studies. Research on environmental issues of the former industrial area as well as high human resources is the main research topic at LTCP. The presence of an updated laboratory for the environmental measurements should also be considered, because, not only in LTCP but also in the larger Lavreotiki region when demanded, basic environmental measurements are conducted (Figure 4.7) (URL-26).

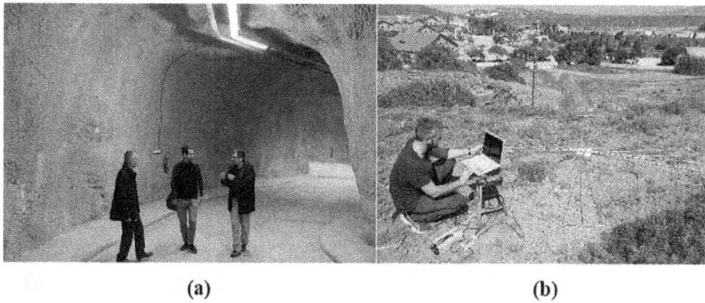

(a)   (b)

Figure 4.7 Research projects at LTCP (URL-26).

4.6. Reuse and Restoration of Abandoned Mining Plants

4.6.1. Landscape of Lavrion and LTCP

According to the European Landscape Convention, also known as the Florence Convention, "Landscape, as perceived by people, is the action and means a field that is the result of human factors and their interaction" (Çiner, 2021). The study of any landscape may require a flexible evaluation of different disciplines such as geography, history, architecture, archaeology, geology, philosophy, ethnography, and several other possible academic fields together. Landscape can be perceived as a stage only for human life. Also, it can be considered the most determinant factor determining human actions and compliance mechanisms and simultaneously being shaped by them. The concepts memory, identity, social order, and reuse are not associated with the understanding of landscape, neither as natural nor as completely cultural (Knapp and Ashmore, 1999). The Lavreotiki region (with an area of ~200 km²) is located at the southeastern end of Attica. This region is dated simultaneously with the extraction of rich mineral resources in its lower layer. The mining and metallurgy sector covers a supply zone, transportation systems (such as roads, ports, railways), capital, equipment manufacturers and distributors, political regulations, laws and regulations, and of course a human community (workers, technical personnel and engineers) capable of performing these laborious activities and enterprises that are labor intensive and complex (Pfaffenberger, 1998). These are actually a list of prerequisites for this industry. With some necessary updates, these prerequisites are valid for both old and modern times, and these conditions determine the success and lifetime of the mines. However, when natural resources are consumed, all mining works are halted and only abandoned mine areas are left (Yıldız et al., 2016; Yıldız, 2020). These areas are becoming industrial heritage areas in nature without human intervention. The landscape itself becomes an industrial/cultural heritage site with a unique identity that reveals the technological, industrial, economic, and social history of the place (Belavilas and Papastefanaki, 2009). The cultural landscape involves the (anthropogenic) effects and components caused by the natural conditions, which have both material and non-material values embedded in the memory, subjected to multidisciplinary evaluations, and affected by humankind. Lavreotiki is composed of many small regions that have been shaped by human activity over thousands of years in a process that still functions dynamically today. Today, the landscape of the region southeast of Attica, defined by hills, valleys, and coastline, includes settlements (with traditional or modern styles); agricultural land (especially vineyards and olive groves); vacant land, roads, paths, and other transportation; industrial units; tourist facilities; geological formations; archaeological and historical sites; and protected forest areas or dwarf forest and bushes, all often facing the sea (Terkenli, 1996; Doukellis, 2009; Hatjimichalis, 2010; Kapetanios, 2013). Lavreotiki has a wonderful and attractive landscape. In addition to this, what binds them all together has been the large underground multi-metallic ore deposits of this region from ancient times to the present. Extracting these minerals has shaped the landscape of the region and land use (Kayafa, 2018). Indeed, the landscape of Lavreotiki changed drastically in the 19th and 20th centuries. As a company town, Lavrion was founded on behalf of FMC. Although there are separate housing units for workers and engineers, the town of Lavrion and its suburbs are given an industrial character (Belavilas, 2012). The first industrial railway line in Greece was built in the region and connected the company center to the mine villages and Lavrion to Athens (Dermatis, 2006).

As highlighted, the town of Lavrion, known for its richness of silver since ancient times, houses a significant historic industrial plant. This plant, firstly built by FMC in 1875, covers an area of 512 decares. In addition, within the LTCP area, there are 48 completely renovated buildings covering an area of ~45,000 m^2, mainly hosting research and development (R&D) activities (Fotiou et al., 2011; URL-14). Most of them were built between the years 1875 and 1940. In the development of Lavrion as an industrial city, FMC had a significant role. The closure of FMC caused the people of the town to be unemployed (Chapter 3). This whole area historically tells the development of mining and metallurgy. Many buildings in this area that operated in the early 20th century are exceptional examples of industrial architecture. It is assumed that the buildings, together with their equipment, are preserved in their natural state (Fotiou et al., 2011). Electric power plants (Figure 4.8) and industrial equipment

Figure 4.8 Electric power plant (URL-24).

Figure 4.9 Industrial equipment (URL-24).

Figure 4.10 a) Machine shop, and b) machine buildings with the LMMM (Urban Environment Laboratory, 2009; URL-24).

(Figure 4.9) are some of the protected sites. The Machine Building (Figure 4.10) housing the LMMM is one of the oldest and most famous buildings in Greece (URL-24). Thus, one of the main exhibits of the new museum was foreseen to be the mechanical equipment of the Machine Building of the LTCP. All machines were taken under protection to be functional (Chatzi Rodopoulou, 2020).

4.6.2. Reuse and Restoration Projects for the Buildings in Lavrion

The LTCP plants contained industrial, laboratory, and professional company buildings with high aesthetic and architectural value, most of which were built between 1875 and 1940. Mineral processing plants were also included, as well as FMC management and auxiliary buildings. The LTCP area housed industrial activities until 1988. During its 120-year activity period, mining/metallurgy plants were subjected to various reuse, renovation, and additions to meet the demands of the technological evolution (URL-13). The transformed buildings of FMC now house various commercial, technological, and cultural activities (Chatzi Rodopoulou, 2020) (Figure 4.11). In the 1990s, LTCP, with 41 buildings and a 45,000-m^2 construction area, was declared a protected area by the Greek Ministry of Culture, and it covered an area of 250,000 m^2 (Touliatos

Figure 4.11 FMC's converted and restored buildings (URL-22).

ve Efesiou, 2010; URL-6). Thus, it was accepted that LTCP was a national heritage protected area as a whole, as a unique work of industrial archaeology and architecture (URL-13). LTCP Board Member Prof. N. Belavilas made the following statement on the matter:

> The new program was inspired by the desire to continue the technology, production, and innovative ideas already existing in the field. The cultural part emerged with the idea that the identity of the place could not be separated from the concepts monument and culture.
>
> (Chatzi Rodopoulou, 2020)

Today, after the end of mining activities, the Lavrion region maintains its historical importance as well as its magnificent beauty. The industrial buildings built in the 19th and early 20th centuries, which are described as historical, protected works by the Greek Ministry of Culture, constitute important works of industrial and metallurgical technology of the past and a large part of Greek industrial archeology (URL-11). Lavrion buildings in their hundred-year lifetime, have successfully adjusted to big functional changes such as the typology of building blocks (within the scope of examining and classifying archeological ruins). Today, restoring and reusing the buildings is foreseen. The future of buildings that could last another century should not be defined by certain uses that we cannot predict the period of (Fotiou et al., 2011). For this reason, projects have been developed for reuse.

FMC plants, built in 1876, are located behind a hill to the north of the Kyprianos neighborhood. During the aforementioned years, the industrial plants on this scale were not only rare in Greece but also in Europe. Forty-one metal processing units were at the Lavrion site, which saw significant technological developments in mining and metallurgy. The best part of the buildings and equipment are still extant today thanks to the attempts of NTUA to establish

Figure 4.12 FMC buildings and plants (URL-22).

LTCP (Figure 4.12). The main idea in protecting the LTCP plants was to protect the creative and innovative spirit of FMC reflected in these stable industrial buildings. The plans and architecture of this industrial plant that is open to expansion were both protected and developed (URL-15).

4.6.3. Restoration Plan and Master Plan of LTCP

In LTCP, a series of buildings were restored based on the plans by way of preserving the original character and providing modern infrastructure and conveniences[4] (URL-3). In this restoration, an activity plan, which aimed at the necessities of the local community and labor market based on a general survey conducted by NTUA, was conducted. In parallel with this, a database was set up with topographical, structural, and photographic documentation for all existing buildings. The archives of the FMC (Figure 4.13) were found and preserved. The construction and installation of the LMMM among the historic buildings of the LTCP, as well as the preservation and organization of Lavrion's rich historical industrial archives (1864–1981), aimed to significantly strengthen and increase the cultural structure of the LTCP (URL-23). As well as models of LTCP buildings, a new map of building blocks was created. With these documents, the entire LTCP facility was designed under the management of the Faculty of Architecture team (Fotiou et al., 2011):

- A new backbone and circulation for LTCP and a new and different structure for open areas were foreseen to be created. In this way, all the units of LTCP were joint and held together (Figure 4.14).
- A certain use scenario for each building and the open area was suggested.
- The possibilities of future buildings and outbuildings on the existing ruins of the past were also considered.
- Within the scope of the policy of the design, minimum intervention for the identity of the place (Figure 4.15) was intended.

(a)

(b)

Figure 4.13 a) (In FMC archives) general layout of Lavrion plants, 1876, and b) post-restoration outlook of LTCP in 2009 (Fotiou et al., 2011).

Figure 4.14 Drone image of LTCP field today (URL-23)

Figure 4.15 A section of houses on the front in Lavrion (URL-22).

A process was developed within the following scope to draw a master plan (Fotiou et al., 2011):

1. Evaluation of the three-dimensional spatial and historical process of buildings in time.
2. LTCP from a landscape and industrial point of view.
3. LTCP's function as a science and cultural park.

As can be understood from the archives and the place itself, LTCP developed in three periods according to its function, activities, and changes in production: a) 1875–1905, b) 1906–1929, and c) 1930–1989. At the same time, this area was explained as three-dimensional, with different typologies, morphology, and structure. LTCP, as a landscape, was produced by two collateral elements: the topography of the sloping hills; railways; and all the elements that create buildings, streets, and open spaces parallel to the slopes created by the mining and metallurgical process were evaluated depending on the capacity of the three units mentioned previously to accommodate new functions, their location in the LTCP and their structure. To this end, each building and its land were evaluated according to the typology and measurements for reusing and hosting opportunities. Various scenarios were discussed together to reach a final result. Eight activity segments were created from various scenarios:

a) research and scientific production; b) renewal and transfer of technology; c) quality controls; d) service network; e) vocational training, long-lasting training, and convention facilities; f) food and hospitality; g) documentation and display of the cultural heritage of the building block; and h) cultural and artistic activities (Fotiou et al., 2011).

4.6.4.   *Restoration and Reuse of Flotation Plant*

The flotation unit ("La Flotation") is located at the center of the former core of the company's industrial zone (Figure 4.16). This building was constructed as a mineral processing unit for hydromechanic ore concentration. It functioned in this way until 1932–1934. Then, until the closure of the plant, it housed a new flotation method for the concentration of low-grade sulfide ore. With its heavy stone walls, transverse light wooden shutters, wooden mezzanines, and tiled wooden pitched roofs, this building was organized on levels of various heights. Designing the existing industrial buildings brought forward some points of order: a) In cases where a historical building has undergone significant changes such as enlargement, demolition, and renovation over time, according to which criteria are the historical periods of this building to be preserved and restored? b) What is the effect of the suggested use on the design process of the future function, since it will not become anything but an example in the future life of the building? These questions were detailed during the design of a highly significant building in terms of its size and history, known as La Flotation. Since the flotation building was exposed to significant changes – such as the prolongation of the period, destruction,

Figure 4.16 Three-dimensional but perspectiveless view of flotation building (NTUA research program) (URL-22).

and modification – directly related to its function in time, the question as to in which historic period the building was restored was brought forward. These changes were transferred onto the printed building image and preserved through drawings, texts, memoirs, or simply the collective memory of the local community. These images were used as data in the restoration process, which has the task of reflecting the future changes; the age; the memory of the building's life; and its existence as space, type, and historical composition. To identify, understand, and present the historical stages of the building, systematic and documented historical research was conducted based on the company files. Protected drawings present the changes in the production methods and the technological equipment, changes in the size of the building, and the characteristic stages of the metallurgical development in Greece. The methods for the reuse and restoration of La Flotation are as follows: a) The design was processed to enable use of the building by different users and various activities. For this reason, two alternative scenarios were applied to decide the minimum common elements fulfilling the needs. According to the first one, La Flotation was going to be home to a small convention center and to a multifunctional center containing FMC archives and exhibition halls to disseminate the history of the buildings to the activities of LTCP and at the same time to support LTCP. According to the second one, the building was going to function partially as a renewable energy center and partially as an administrative, entertainment, and cultural center. In both situations, the minimum common factors were as follows: 1) modernization of transportation using new technology (stairs/elevators); 2) Modernization of accessibility to services and information (with labeled and newly designed access points); 3) new electricity and water supply, sewerage, ventilation, and heating networks; 4) addition of cleaning and service facilities (toilet units, warehouse, dining hall); and 5) presentation of the structural capacity of the building together with the architectural elements. b) It was planned to use a new design method for new interventions. Several new interventions were designed to indicate the building's topography, typology, and history (Fotiou et al., 2011). Images of the flotation building after restoration are presented in Figure 4.17.

Figure 4.17 Photos of the flotation building after restoration (Fotiou et al., 2011).

Figure 4.18 Buildings in Lavrion (URL-15).

4.6.5. *Restored Buildings in Lavrion*

The architectural and housing development of Lavrion was considerably affected by the mining and metallurgy industries that were active once. While the FMC operated the mines in Lavreotiki, the industrial areas were next to the modern town of Lavrion. Industrial buildings and plants were built to meet the needs of the companies operating in the region as well as the accommodation needs of their personnel (Figure 4.18). In addition to this, the following economic growth encouraged the construction of various specific buildings to meet the religious, cultural, and other needs of a developing community. Thus, the modern Lavrion character, determined with closer historic interventions, was blended with factors reflecting the living industrial past. Today, old plants are considered worthy of preservation, and a new vitality was brought to these plants. About 40% of old plants have been restored, and the soil in the surrounding areas has been rehabilitated. Thanks to all these activities, the old FMC site and buildings have become a vibrant technological and cultural park (Panagopoulos, 2008; URL-15).

In the 19th century, many buildings that were popular at that time in the neoclassical style were built, and many of them have survived to this day. A representative sample could be the Euterpe building located on the edge of the palm tree park. Euterpe's neoclassical building, with portraits of Sophocles, Orpheus, Mozart, and Verdi on its worn ceilings, harmoniously coexists with other buildings, including the two impressive churches of Kyprianos (Figure 4.19): Evagelistria Orthodox Church and St. Barbara Catholic Church (Figure 4.20). Another typical example of neoclassical architecture is the building of the Philomouses (music lovers) Association, built by the Greek Metallurgical Company to house the city's Philharmonic Orchestra. Located in the central square, this building, together with Euterpe, formed the second cultural center of Lavrion, proving active participation by its inhabitants in the arts (URL-15). Also, this building served as a performance area for theatre and dancing. The impact of neoclassicism is also obvious in the old municipality building located in the center of the city. It is a simple cubic building interrupted by a prominent balcony on its façade, where Greek Prime Minister Eleftherios Venizelos addressed the residents and miners of Lavrion in 1929.

Figure 4.19 Church of Agia Paraskevi (URL-15).

Figure 4.20 St. Barbara's church (URL-22).

Figure 4.21 Buildings in Lavrion, some of which have been restored (URL-13; URL-15).

A few meters away is the city's famous bazaar (known as Psaradika). The construction in the neoclassic style, which was built in 1885 by the Greek Metallurgical Company, still constitutes the gastronomical heart of Lavrion, with taverna and restaurants as well as fish and meat shops. The LTCP area and some other buildings restored in Lavrion are presented in Figure 4.21.

4.6.6.   *Lavrion Port and Industrial Activities in Lavrion*

The establishment of mining companies in the 19th century led to the constitution of an important export center. It is quite remarkable that the capacity utilization of Lavrion Port increased from 40,000 tons in 1860 to 450,000 tons in 1899.[5] The large iron dock of FMC, as well as the stone sea quay of the Greek Metallurgical Company, were unique examples of maritime work at that time, with metal bridges fitted with cranes to load the products of local metallurgical companies onto ships. There were many other industrial buildings constructed to support the active port (these former industrial buildings now house the private enterprises or public services). For Lavrion Port to be conserved, the port and industrial buildings have gone through comprehensive restoration works. Right next to the port, French Pier and abandoned houses of Lavrion from the FMC era overlook anchored merchant ships, passenger ships, fishing boats, and sailing boats. Today, Lavrion Port, with its privileged position near the International Athens Airport and the Cyclades, contributes to the increasing tourism and marine opportunities (Figure 4.22a). The Lavrion Port (Figure 4.22b), promising and bearing the signs of its magnificent, industrialized past, is known for its central role in the tourism and cultural life of Greece (URL-15). However, the dock loading piers, which are closed now, were neglected. They could become interesting centers of attraction if restored comprehensively (Periferakis et al., 2019). A larger area of Lavreotiki apart from mining constitutes a part of the sea tourism area of Attica and Cyclades that covers 45% of tourist demand (URL-12). According to the research conducted in 2001, the economic activities in the Lavrion buildings are presented in Figure 4.23. Based on this, it is active in areas such as commerce (10%), hotel-restaurant (7%), transportation

(a)   (b)

Figure 4.22 a) Lavrion boat docking (Url 15), and b) Lavrion port (Url 15).

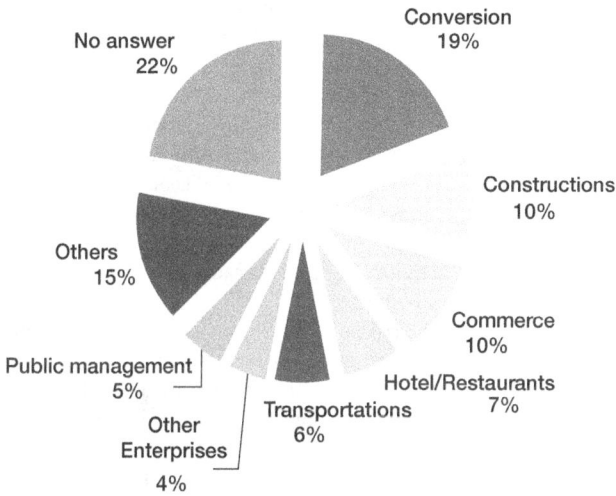

Figure 4.23 Economic activities in Lavrion buildings (URL-12).

(6%), other enterprises (4%), and public administration (5%), even though there are buildings being transformed (19%) and constructed (10%). Other uses of the buildings are stated as 15%.

4.7. Rehabilitation and Environmental Projects and Activities in Lavrion

4.7.1. Rehabilitation Projects in Lavrion

Intensive mining and metallurgical activities caused the deterioration of the environment in Lavrion. Disposal of mine and metallurgical wastes in Thorikos Bay in the past caused pollution here through the effect of acid mine drainage and heavy metals. The natural landscape and rich forest vegetation in Lavrion were swiftly spoiled. Bogs and coasts were covered with solid waste; verdant forests were burned by arson. The smoke coming from open pits covered a larger area of the city for kilometers (Panagopoulos et al., 2007; URL-3; URL-22). Despite this, as a result of the decrease in intensive mining activities in the years after the Second World War, Lavrion and its surroundings became forestry again (Photos-Jones and Ellis Jones, 1994). A significant part of Lavreotiki was declared Sounion National Park in 1974. The decision to declare Sounion National Park was probably made considering the historical importance of the region, the dense population of Attica, and the need for open spaces and forests. This park is the smallest national park in Greece. Most of the forest is covered with Aleppo pines (*Pinus halepensis*). There is a wide variety of plant species and wildflowers. Some of these are protected by CITES. The fauna of the park consist of various reptiles; mammals such as hedgehogs, rabbits, foxes, weasels, and badgers; and most of the migratory bird species that use the park as a place to stay (Vallianatou, 2012; Kayafa, 2018). Despite its beautiful nature, the emerging oxidized waste of sulfur ores that are the wastes of mineral processing activity (~1,000 m^3 in size) is in Lavrion as a relatively small stack. These are siliceous and hard materials (FeS_2) concentrated in various areas around the former mining facilities of Lavrion (current LTCP area). Sulfur-containing substances discharged after mineral processing activities are toxic because of partial oxidation. Due to these toxic elements, in both sea and terrestrial area, a residual pollution source emerges. These have risks of affecting human health significantly. For this reason, these elements pose an environmental threat to locals in Lavrion (Janikian and Theodossiou, 2009). A study focused on determining the characterization of air quality in the abandoned mine site and the pollution source in Lavrion was conducted. In the last 3,000 years, the severe soil pollution resulting from intensive mining and metallurgical activities occurring in a larger area than metallurgy plants has emerged as a big problem. Sediments that are rich in terms of heavy and toxic metals and contaminated on the surface cause particulate matter to be conducted in the air since they can be easily eroded by the wind (Protonotarios et al., 2002). To make this environmental hazard nonhazardous and keep it under control, environmental protection and rehabilitation projects were conducted in Lavrion. These projects were developed in harmony with the other projects within the scope of LTCP.

For the sustainable rehabilitation of the disposal of area of an old mine waste covering an area of 10 ha in Thoricos Gulf in Lavrion, the fundamental principles of risk management and project design were applied. This rehabilitation project aimed to return this coastal area from a polluted area to a suitable recreational area. Apart from the data on hazards obtained from the comprehensive studies to identify field data, a basis is provided to determine the related environmental pollution risk. To remove this environmental risk, the option of minimizing the risk was chosen, and a rehabilitation plan according

Figure 4.24 Covering of waste with clay layers and HDPE liners (URL-20).

to environmental criteria was determined. This plan consists of in situ neutralization, stabilization, and barrier coverage of potentially acid-producing wastes and contaminated soils with low geochemical transmissivity. In areas where high-sulfide wastes are generated from active acid-producing resources, additional measures were especially applied, including covering the wastes with high-density polyethylene (HDPE) liners over clay layers (Figure 4.24). This integrated rehabilitation plan also includes the construction of a rock embankment set along the coastline and drainage channels to protect the recovered area from sea waves and water runoff erosion. For the design of this risk management plan, the expertise of a multidisciplinary team of geologists; hydrologists; mining, exploration, chemical, civil, coastal, and environmental engineers; acid mine drainage specialists; and ecologists was used constructively. The different phases of the project were carefully designed to allow timely implementation within the scope of environmental risk management within the specified time and budget. The Thoricos Gulf rehabilitation project was considered one of the most challenging environmental projects conducted by the Greek public sector for the rehabilitation of a historic mine site (Panagopoulos et al., 2007).

All necessary evaluations and supportive projects according to the EU and Greek legislation were conducted, all necessary health and safety measures were taken, and their applications were realized to conduct this project. In the approval process of all projects and permits, public opinion and management approval were considered according to the national legal framework. The environmental impact assessment of the landfill was approved by the local municipal board. Also, with the help of the LTCP Laboratory and other NTUA laboratories, consistent and attentive monitoring was realized for the planned construction and operation of projects. Monitoring is not limited to applying environmental conditions regarding certain projects. However, the rehabilitation project expands to the application of environmental protection, additional measurement, and tests as a sample project for designing similar projects and in terms of safety. The specific environmental response resolved a major environmental pollution problem that represents the main source of pollution in a larger area than LTCP. On the other hand, the project represents a significant specialty resource to take care of similar environmental problems in other regions of Greece with similar problems, starting with the Lavreotiki region. On behalf of the NTUA, the scientific committee responsible for the project was in touch with the people living in the large area around Lavrion or their representatives to inform them more about the project. Additionally, an integrated project continues to be applied for project results to become prevalent (URL-13).

Thus, NTUA AMDC, with the Competitiveness Project under the leadership of the Ministry of Development with a total budget of 7,800,000 €, it undertook the Soil Rehabilitation and Regulation Project in LTCP as part of environmental rehabilitation projects to meet a vital environmental responsibility in a highly degraded site such as LTCP. This rehabilitation project aimed to apply environmental projects in the region (URL-13; URL-20). More specifically, these projects contain the following (URL-13):

- The activities of digging in a specifically created place within the LTCP area, moving and safely accumulating the contaminated soils obtained from different locations within the borders of LTCP, were conducted. This project involves building a hazardous waste repository field designed according to the strictest technical and environmental specifications. In the regular repository area, contaminated soils posing a danger to both human health and the ecosystem have been accumulated safely.
- An embankment was created from all removed contaminated soils and these soils were replaced with clean ones. After the previous procedures were completed, the repository area was permanently closed and sealed. All environmental parameters here are consistently monitored.
- For technical and environmental safety reasons, an underground area was created within the boundaries of the LTCP in order not to mix the soils accumulated in the landfill with the hazardous waste. Similar to the contaminated soils mentioned previously, this waste was a by-product of former mining-metallurgical activities developed in the LTCP and could be found in various locations within the LTCP area. The waste mainly consists of arsenic compounds and is intended to be collected in special drums and bags. This allows is to be accumulated safely and for a long time in the underground area. It was emphasized that only the waste from LTCP would be transferred to the underground repository area and in no way would waste from other sources would be accepted.

A laboratory specializing in environmental measurements has been established in LTCP, which aims to provide the aforementioned large-scale technical projects environmentally. This laboratory monitors and records the data of all environmental parameters during both construction and operation by performing special analyses and carrying out on-site measurements regarding all environmental aspects (earth, water, and air) (URL-13). The Soil Rehabilitation Project

Report clearly states that no heavy metal concentrations have been detected. The rehabilitation plan carries specific scientific and educational value since it is a unique case study of environmental rehabilitation of 25 decares loaded with 150,000 tons of low sulfur, in addition to its practical importance (URL-31).

4.7.2. Features of Underground Waste Repository in LTCP

Despite the fast restoration of LTCP buildings, only the most critical areas were focused on to reduce the land rehabilitation work emergency risks. While the degree of contamination varied due to the successive transfer and accumulation of waste over the life of the metallurgical plant, most of the floor was covered with waste and slugs. Chemical analysis of the soil samples in the LTCP area revealed high levels of lead, arsenic, zinc, cadmium, and copper (Karachaliou and Kaliampakos, 2005). See (Benardos and Kaliampakos, 2013). LTCP rehabilitation project includes storing ~115,000 tons of contaminated soil by removing hazards and constructing an underground repository (Figure 4.25) to collect the most hazardous waste in the LTCP site (Kaliampakos et al., 2007).

The underground repository was specifically designed to store ~5000 tons of hazardous waste (with the option of monitoring and recoverability) containing high concentrations of heavy metals and toxic metalloids such as arsenic, lead, cadmium, and zinc (Kaliampakos et al., 2009). In the decision making concerning such an in situ underground storage arrangement, prominent considerations were as follows (Benardos and Kaliampakos, 2013):

- Complete containment of hazardous wastes (Kaliampakos and Benardos, 2009a, 2009b).
- A cost-effective storage plan that can serve as an example in similar situations around the world.

(a)   (b)

Figure 4.25 a) Construction works for underground storage in LTCP (URL-20), and b) underground storage adit in LTCP (URL-22).

Since there is no central underground or surface hazardous waste disposal repository, the only available other choice was to transfer the waste to another storage area that was far away. However, this alternative would require a complex transfer procedure and a very high disposal cost. Thanks to the development of the LTCP underground repository, storage with relevant costs compared to the cost stated for the alternative became possible. This alternative was directly compared to transferring waste that are the other available healthy option in terms of the environment to underground storage or specific storage areas abroad to evaluate the cost benefits obtained from the development of LTCP underground storage. In this option, the cost of transportation and disposal changed according to the characteristics of the waste and varied between 1,500 and 3,000 € per ton. The total construction cost of the LTCP underground repository was €2.8 million. Considering the maximum storage capacity of the LTCP underground repository, it was calculated to cost around €560 per ton, which is at least three times less than the cost of transport and disposal abroad. In addition, support from local people for the project was provided since LTCP would keep the hazardous waste and provide significant improvement in the general environmental conditions of the region. During all design and construction processes, continuous communication was maintained between LTCP management and the local people residing close to LTCP. Many meetings with and special presentations to the local people were conducted to understand the aim and development process of the project (Figure 4.26). The key issue to gaining public acceptance for the

Figure 4.26 Informing the public in the underground hazardous waste storage at LTCP (URL-20).

Figure 4.27 Photos of LTCP storage ready for hazardous waste storage (Benardos and Kaliampakos, 2013).

development of the underground repository was the assurance that the waste would originate only from the LTCP area and that no waste from anywhere else would be stored in this underground repository. This assurance showed that by the end of the project, the quality of living conditions of the people of the region would be significantly improved (Benardos and Kaliampakos, 2013).

The underground waste repository is located under a small hill in the northeast part of the LTCP field (Koumantakis et al., 1999). Generally, geotechnical conditions on the construction level were good: The rock mass can be characterized as medium-good quality, with rock mass rating (RMR) values ranging from 52 to 67. The design of the underground repository (Figure 4.27) was made using an underground mining method, the room and pillar method (Benardos and Kaliampakos, 2006). The usable area of the facility is ~1,900 m², and it was built on a total area of 2,475 m². Access to the surface consists of an adit that enables waste transportation and general access to the storage and is 10% inclined and a shaft with 35-meter depth to be used for safety and ventilation (Benardos and Kaliampakos, 2013).

4.8.   Education and Other Activities in Lavrion

LTCP's recently renovated plants include an area of ~1,280 m² that can house other professional or social activities as well as conferences, seminars, exhibitions, and movie and commercial shootings (URL-3). Also, the LTCP field, with its ancient mine tunnels, offers its visitors a unique representation of mining-metallurgy technology starting from the ancient ages to the 21st century (Efesiou, 2011).

LTCP is one of the most visited places by students, with over 7,000 annually both domestic and abroad, as well as the members of scientific institutions, international business committees, graduates and studying students (URL-26). Indeed, the students in Greece were asked whether they had visited the industrial heritage areas in Greece. Eleven options were given, and more than one answer was possible. Among these, Piraeus Maritime Tradition Museum was the most visited (31%). Next (with 24%) was the LTCP in the city of Attica (Polyxen and Spyridon).

LTCP offers training services independently or through collaborations at every level from elementary to upper-level education activities for young scientists, researchers, and professionals in specially designed places. Conferences, seminars, and workshops arranged by NTUA or other universities or educational institutions are held here. Additionally, within the LTCP field, there are two educational institutions, EECL and HIEM. The first one is a training center, supported by the municipality of Lavreotiki, which, as an institution of the Greek Ministry of Education, provides training programs for schools and training seminars for teachers (URL-29). Like EECL, about 50 centers are active throughout Greece. In these centers, teachers train students on subjects related to their branches. It is a place-based education (Papadimitriou, 2012) that uses local examples to promote environmental and social responsibilities. HIEM, on the other hand, is a non-governmental organization that is non-profit making and operates under the auspices of Lavreotiki municipality (URL-29). The Attica region is one of the partners of the project and participates in almost all planned activities. Together with two other Greek partner companies (BIC of Attica and LTCP), it focuses mainly on education-leisure thematic areas (URL-3). Thus, a large number of educational activities are conducted in LTCP both within the framework of national education and in parallel with it (Figure 4.28). Within this scope, LTCP includes/organizes or provides the following (URL-25):

- Scientific conferences, seminars, and workshops for scientists, university and other school students, and the public.
- Temporary recruitment of students to complete their academic internship (Atlas system).
- Support of undergraduate and graduate students by providing materials, areas, or guidance for their projects.
- Mediation for research and educational collaborations between LTCP laboratories or companies and students or young scientists.

Figure 4.28 Educational activities in LTCP buildings (URL-29).

- Activities to get to know the history of the LTCP area and its larger environment through guided tours in groups.
- Educational programs, educational games, competitions, and other events.
- Student projects that will be an indicator for all these activities.

The many features stated in the study are the greatest proof that Lavrion can offer multifarious educational activities. Features of geological, mineralogical, and ore formation processes and conditions at individual locations are available at each site and are educational. For educational purposes, lectures and training on Lavrion geology, mineralogy, and ore deposit formation are among the most preferred for students in many Greek universities. As a part of the lectures for undergraduates and graduates in the Lavrion region, every year various field trips are arranged. In addition, some undergraduate, postgraduate, and doctoral theses contain research conducted on the Lavrion deposit. For instance, thanks to some of this research, considerable information about the geological and archaeological restrictions has been developed (Voudouris et al., 2021).

Students visiting LTCP vary from kindergarten to the university level. These students can participate in a series of training programs that last for 3–5 hours. Also, LTCP organizes seminars, workshops for teachers' vocational training, and other environmental activities that promote sustainable development for the benefit of the local community. Depending on the age of the students, different training tools and programs are applied and different aspects of Lavrion history are taught (Voudouris et al., 2021). Lavreotiki is a discovery area that is generally interesting and entertaining in environmental and cultural training. It is in line with modern ideas of disseminating knowledge beyond the experts to the public, who choose the future of sustainable development (Papadimitriou, 2012). Lavreotiki is considered an ideal setting for raising environmental and cultural awareness because it provides beneficial case studies on a series of subjects on the matter of sustainable development. Education programs from LTCP involve the flora and fauna of Sounion National Park, ancient mining and metallurgy history, and its technical operation. Also, it covers recent mining history and its effects on the environment and society, the geology of Lavreotiki, the concept of geodiversity, human interventions, the concept of protected areas in Sounion National Park, various aspects of contemporary life in Lavrion, and local history. In any case, the aim is to use cutting-edge teaching approaches where students can watch, walk, observe, think, explore, recognize, connect, and dream by providing first-hand experiences while visiting the industrial heritage site at LTCP. Seeing and touching influence students from every age. Thus, a different type of learning is formed, which focuses on visual content and problem solving and is based on less fixed traditional information transfer (Voudouris et al., 2021).

LTCP is an ideal place for conferences, exhibitions, cultural activities, and seminars. The region's rich history, the wealth of monuments, and great industrial tradition make LTCP a major attraction for visitors. Also, LTCP's proximity to central Athens and easy access to the Athens International Airport, Eleftherios Venizelos, increase its suitability for conferences and other events (URL-3; URL-28). A visit to LTCP does not only offer a tour of LTCP but also participation in the various cultural activities housed by LTCP. Especially in summer, participation is available in the theater performances and musical concerts arranged there. In addition, LTCP is an ideal place for a family trip. The Environmental Education Centre of Lavreotiki and the Handicraft and Industrial Education Museum run programs for children and youth. Through learning by doing, children can have the opportunity to become familiar with mining/metallurgical production methods, tools, and machinery or meet former mineworkers who share their experience working at FMC. In collaboration with the Archaeological Museum, there are workshops regarding the ancient mining processes or minerals, occupations, and clothes in the ancient age. A visit to LTCP can be a fun and educational experience for the whole family. Museums, buildings of historical and architectural value, and industrial monuments scattered around the city invite guests to discover the fascinating history of this small town. This vibrant sea town also offers extensive opportunities for activities such as relaxation, good food, entertainment, walking, or swimming (URL-15).

LTCP is an attractive venue for numerous cultural events. Music concerts and scientific/artistic exhibitions, theater and dance performances, film screenings, conferences, seminars, and corporate and other events have been held in the open-air arena, conference, and exhibition halls during the 20–25 years of operating of LTCP. The recently renovated plants of LTCP are 1,280 m² in size and provide technological equipment and complete supply for the best organization of any event. Also, at the park, there is an open-air theatre with 700–1,200 seats, extended gardens, and halls housing activities in addition to a parking lot. The Pharmacy Hall is the only place in a stone building with a wooden roof made of Byzantine tiles. It is fully equipped with a 4 × 4 projection screen, video projector, and sound equipment. This hall, which is ideal for lectures, workshops, and meetings, has an area of 110 m² and a capacity of 80 people in a theater arrangement (URL-28).

Right after the mining-metallurgy activities ended at the former FMC site, this abandoned mine area became a "natural" setting for series, film, television, and commercial productions. Many movie shoots with major award-winning directors, a series of elite actors, and teams have been successfully conducted (Figure 4.29). What makes the area unique are the various cinematic frames in which different architectural scenes from historic houses to natural habitats are presented (Figure 4.30). The LTCP's extensive park and installation space, and its proximity to favorite movie theaters and many interesting places of Lavreotiki, increase the artistic popularity of this area (URL-30).

Figure 4.29 Filming in Lavrion (URL-30).

Figure 4.30 Movie scenes shot in Lavrion town (URL-30).

Notes

1 One of the most characteristic examples of such a solution in the Attica region is turning the former gas plant into a municipality cultural center by reusing it by restoration (URL-3; URL-22).
2 In Europe, regional science groups, seen as essential drive forces for competitiveness, innovation, and regional development, are encouraged both by measures at the national/regional level and by means of support at the EU level, particularly by the 7th EU Research Framework Programme, the Competitiveness and Innovation Programme and the Cohesion Policy Funds (URL-9).
3 The MED regions are: Attica, Lazio, Sicily, Calabria, Lombardia, Algarve and Valencia (URL-9).
4 These buildings can be leased from technological companies. In parallel with this, it was stated that (URL-3) restoration had yet to be conducted and there were still opportunities for interested investors to offset the renting costs and cover the expenses of restoration.
5 This high level of commercial activity also led to the construction of some of the most beautiful industrial buildings, such as the clock building (it was necessary so that workers who could not afford wristwatches knew when to start and finish their work.) (URL-15).

5 Industrial and Mining Heritage and Geoheritage Characteristics of the Mine Site in Lavrion and Around It

As mentioned in Chapter 3, starting from the Classical age, Lavreotiki became the main silver and lead supplier of the city-state of Athens. On this matter, literary resources give dates as well: 483 BC is the year the Athenians discovered a rich vein of silver. Lavrion is described as a fertile mining region for Athens, especially in the 5th and 4th centuries BC. The abundant mining remains in the region demonstrate this. Lavrion holds remains that confirm that galenite and cerusite were extracted or processed in ancient times to produce silver and lead. These include mine shafts and adits, very deep inclined shafts, mines, water reservoirs, several furnaces for mineral processing, quarries, and towers, as well as residences, farms, tombs, shrines, and a theater that bears witness to a permanent population. All these have been determined as a result of archaeological evidence/ruins (Photos-Jones and Ellis Jones, 1994; Saliora-Oikonomakou, 2004; Apostolopoulou-Kakavoyanni, 2008; Kapetanios, 2013; Kakavoyannis and Koursoumis, 2013; Kayafa, 2018).

The Lavreotiki region is located in the southeast of Attica, 55 km from Athens, and covers an area of about 200 km². Its morphology is rough and semi-mountainous. Its highest altitude is at Megalo Ribari (372 m). The area is dry and partially covered with pine trees and crosses many valleys. Access here is accomplished through asphalt roads that are relatively easy to travel. The region's history, scenery, natural beauty, proximity to Athens, and hotel infrastructure make the Lavreotiki peninsula an ideal destination for short or longer visits. The best way to get to know the rich history of Lavrion is by wandering around the city. Neoclassical buildings (Figure 4.18 and Figure 4.21); picturesque streets; Town Hall; the J.B. Serpieri Monument; the Mineralogy and Archeology Museum (Figure 5.25); the churches of Agia Paraskevi (Figure 4.19) and Evangelistria, built in 1885–87; the Old Market; and the French Ore Loading Port and dock (Figure 4.22), built in 1888, are must-sees during a town visit (Janikian and Theodossiou, 2009).

The geological landscape of Lavreotiki includes many places where one can witness the tectonic and petrological history of the region (Kayafa, 2018). The remarkable geology of Lavreotiki can be seen in a series of geosites

DOI: 10.4324/9781003408437-5

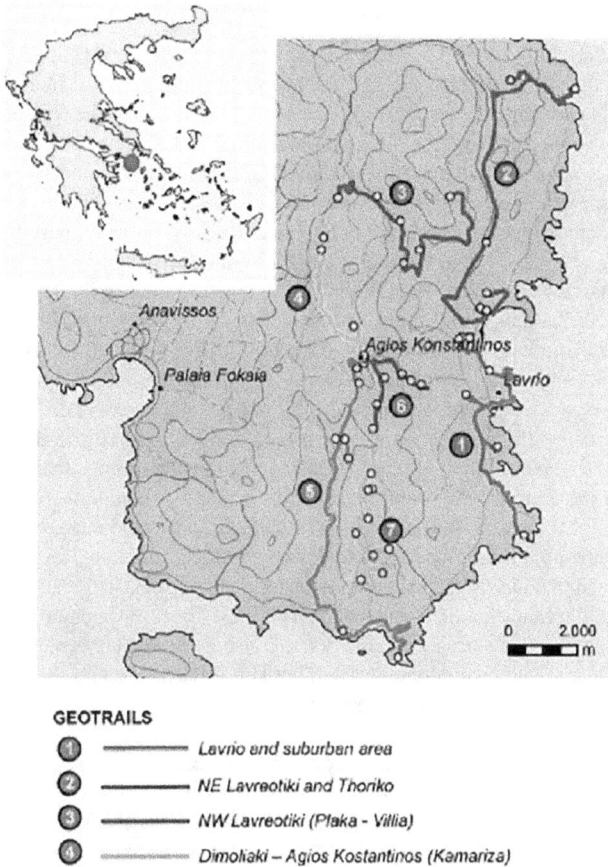

GEOTRAILS

① ———— *Lavrio and suburban area*

② ———— *NE Lavreotiki and Thoriko*

③ ———— *NW Lavreotiki (Plaka - Villia)*

④ ········· *Dimoliaki – Agios Kostantinos (Kamariza)*

Figure 5.1 Locations and morphology of Lavrion (Janikian and Theodossiou, 2009).

recently identified by the IGME as part of the Geotrails series in Greece (Jani-kian and Theodossiou, 2009; Kayafa, 2018). Within this scope, a tour area map suggested for research of industrial heritage areas in Lavrion is presented in Figure 5.1.[1] For certain geological, mineralogical, and metallurgical areas, a total of 47 geosites were chosen and registered. Geosites were divided into areas and seven parkours were determined (Janikian and Theodossiou, 2009). Lavrion's industrial and mining heritage areas can be visited according to these parkour spots. The main areas in and around Lavrion and their features are explained in the following subsections.

5.1.   Sounion National Park

Sounion National Park is the heart of Lavreotiki, with its features appealing to archeology, paleontology, geology, and botanical sciences and an area of 4000 ha. This park, constructed in 1971, is under the protection of the Natura 2000 Network. The verdant flora of the park contains one of the biggest pinewoods in Attica, multifarious plants of the Mediterranean, and two endemic knap-weed species (*Centaurea laureotica* and *Centaurea attica*). The local fauna include many bird species, reptiles, and mammals, while in the past it was a hunting ground for wild boars and deer. It is said that Otto, the first king of the modern Greek state, founded in 1832, hunted in the region. The diverse karstic formations found in the park have led to the revealing of numerous plant and animal fossils that testify to the existence of now-extinct species. In addition to this, the surprisingly rich geology of the region, and around 260 mineral types – some of which are specific to Lavreotiki – prove the intensive mining and metallurgy activities date back to the Neolithic Age in the region (URL-15). Hence, all of the park areas contain ancient mines, shafts, metal processing plants, houses, and cistern ruins, waiting to be discovered. Sounion National Park, a typical example of a Mediterranean landscape that has indisputable importance for the microclimate of Attica and biological diversity, is an area that has unique natural and historical value that will make visiting worthwhile (Kayafa, 2018; URL-15). With these features, Sounion National Park and the area surrounding Lavrion are also a remarkable center for cultural tourism and entertainment. The Thorikon Ancient Theatre, the Temples of Neptune, and Sounion are also worth visiting, as are the ancient mines and mineral processing plants. LTCP is one of the most important tourist places and contributed to the development of the region, centered in the place where the old FMC general center used to be in 1992 (Chapter 4) (Janikian and Theodossiou, 2009). Within the scope of the industrial heritage of Lavreotiki, Lavreotiki Geopark is also worthy of investigating in northeastern Attica.

In the ruins at the Sounion National Park, monuments of various ages can be seen, as well as large adit openings and high inclined shafts. Here, there are intensely colored rocks (mostly red and yellow tones with iron oxide), interesting geological features, quarries, stone stacks, waste heaps, and abandoned buildings. On the road to Soureza, next to the park, there is a geologically remarkable dent named "The Chaos" that was formed as a result of the collapse of a gigantic cave. This area has geological characteristics with mostly red and yellow iron oxides (Katerinopoulos, 2010; Kayafa, 2018). In the larger area of Lavreotiki, there is only a marble quarry. This quarry was opened in the early 20th century and was intermittently active until the early 1970s. The adit to this quarry starts with an excavation similar to a ditch from the outside and continues with a small slope underground (Janikian and Theodossiou, 2009). Large, fenced areas with properly excavated, generally well-preserved, often inaccessible ruins/antiquities can be found (Kayafa, 2018).

5.2. The Accommodation of Mine Workers in Lavrion

In the mining history of Lavrion, a few years after the modern era started in the 19th century, houses were built around the main mining centers to provide a place for miners to stay. What is left today are some single-story old houses without roofs in Soureza (Janikian and Theodossiou, 2009; URL-6). Dozens of lavatories have been preserved in the Soureza Valley, an archaeological site ~2 km from Agios Konstandinos.

5.3. Old Mine Shafts, Adits, and Railways

Serpieri mine Shaft No.1 was constructed by FMC in 1880. It is 327 meters deep. It remained active until 1973. It is one of the most significant mineral monuments in the region. The building houses the Mineral and Mining Museum of Agios Konstantinos now. Apart from the hoisting tower, the building where the starter engine is held, ore loading bridge, and sections of open-air warehouses, where the ore selection was made, can also be seen. Shaft No. 1 was used to extract the ore and pull the workers up and down in that mining section (Janikian and Theodossiou, 2009; URL-6).

Taking the road from the main square of the municipality, ~400 meters to the south, a short distance to the south of Shaft No.1, a sloped adit (Paron) with a 450-meter length is reached. This adit was built by FMC in the early 20th century and was active until 1973. This was the main entrance of the workers in the larger mining section of Kamariza. It was the biggest mining center in Lavrion. The adit is in good condition and can be visited with a guide and proper equipment. Before entering the adit, there is a janitor's building and two guard posts where the entrance and exit of workers were controlled[2] (Janikian and Theodossiou, 2009).

Since prehistoric times, mining activities in Lavrion have been on the rise. Thanks to this, this region contains a complex and unique technological system for the processing of silver-lead ore. This white metal became a true power source in its time. Also, it became a true source of inspiration for mining technologies. Deep shafts in the region enabled the underground area to be discovered in a vertical plane. These shafts constituted the main access route for necessary tools and potential deeper mining levels. The tools used for excavating hard rocks were particularly innovative in old times (Morin et al., 2020). More than 1,000 shafts have been found in the region, some reaching depths of up to 119 meters. In the larger Lavrion region, especially in the Plaka and Kamariza regions, hundreds of shafts and adit entrances can still be seen (Katerinopoulos, 2010). With these features, the Lavrion mines are the only mines known in the ancient world for connecting the networks of an underground adit to benefit from ore deposits with vertical shafts that exceed the depth of 100 meters (Morin et al., 2020). These mines required complex operational, engineering, and technological skills and a perfect geological understanding. The underground setting is a protected area where old mining activities were well protected. Tool marks and fossilized footprints of old miners contain substantially significant information and proof regarding the old and new mining activities on the rocks (Duchène et al., 2018). The two old mine shafts are close to each other. For this reason, they are called twin shafts. The first one has a cross-section of 1.30 × 1.65 meters, and the second one has a cross-section of 1.25 × 1.25 meters. They are 7 meters apart. Both of them are 100 meters deep and have a common adit at the bottom. It is thought that twin mine shafts were built so that miners could work underground without ventilation problems. These shafts were used to extract the ore taken part in the upper elements of the lower marble of the Kamariza Series (Janikian and Theodossiou, 2009).

Ancient adits have a total length of hundreds of kilometers and were built on six floors, connected by numerous shafts. The gigantic, complex network of these ancient adits appears more impressive when considering that they were dug a few meters every month using hand tools or sometimes fire (Katerinopoulos, 2010). Entrances of some of these shafts and adits are still in rural Keratea (Periferakis et al., 2019).

A few meters away from the ancient mineral processing plant, an old adit with its entrance surrounded by a fence is located. This adit is estimated to be ~100 meters long and to be the first adit opened in the Lavrion Region in the ancient age (URL-6). Due to deterioration, this adit cannot be visited anymore. There is also another adit in the region with ~6 km length. In a larger area, ores and deposits preserved in the region resulted from intensive mining activities. Three ancient adit entrances are close to each other with narrow gaps. The activity in this mine essentially focused on operating the mineralization in the upper layer of the Kamariza Series, in other words, initial contact. The ore extracted from this mine was crushed and milled before processing in the mineral processing plants that are next to it. The last adit here (Adit 80) is the last mine site of the modern mining activity period. This adit was constructed by FMC between the years 1956 and 1958 and used for mine production until 1983. It is 2,350 meters long and essentially was used to benefit from the second mineralization contact (upper marble base with Kamariza). The total sulfide ore production is 530,000 tons, and the average grade is 3% lead, 3% zinc, and 30 gr/ton silver (Janikian and Theodossiou, 2009). The Kamariza Railway was used to transport the produced minerals to the port. In this way, the first industrial railway in Greece was put into action in Lavrion in 1871. This railway, which was originally 10 km long, connected Lavrion Port to the Agios Konstantinos region (Figure 5.2). Later, it was expanded in a way that it would involve the main Lavreotiki region, and its length reached 40 kilometers. In this area, the stone bridge, where the railway tunnel (Figure 3.5) and under which the railway ran, can be observed. The railway line was used to load slags and waste of ancient mining activities (low-grade ores), transfer, and unload (Figure 5.3) (Janikian and Theodossiou, 2009).

Figure 5.2 Kamariza Railway (Dermatis, 2017; URL-22).

Figure 5.3 Transport activities on the railroad (Dermatis, 2017; URL-22).

5.4. Features of Mineral Deposits in Lavrion

The first traces of metal extraction in Lavrion date back to the Late Neolithic and Early Bronze Ages (Spitaels, 1984). Enough evidence was found to show that not only lead and silver but also copper was extracted from the mineral resources in the Bronze Age. This evidence emerged from lead isotope analysis, a method that compares the lead isotope signature of Lavrion ores with that of lead, silver, and copper objects. Many samples from metal pieces of the Bronze Age in Aegean were subjected to lead isotope analysis. According to these analyses, it was determined that Lavrion should be accepted as an important copper resource for the Late Bronze Age (Gale et al., 2007; Stos-Gale,

2014). The copper in the region was essentially produced in the Bronze Age, while the iron was mostly produced in the ancient age (Conophagos, 1980; Kakavoyannis, 2005; Morin and Photiades, 2008).

Approximately 600 minerals, some of which were not metallics and some of which were not economically operatable (or grade), in Lavrion were determined. Generally, economically operatable ores in this region are divided into two main categories: mixed sulfide lead-zinc-iron-copper ores (generally form oxidized, prevalent secondary mineral accumulations) or Fe-Mn ores (Janikian and Theodossiou, 2009). These metal types have been produced in Lavreotiki recently or formerly. In the ancient age, the most important minerals produced in the region were galenite, cerussite, and smithsonite (Marinos and Petrascheck, 1956; Kakavoyannis, 2005). In this way, small-scale open pit mining and mineral processing between the years 4000 and 1000 BC focused on silver, lead, iron, and copper on the highly oxidized mineralization related to the initial touch. Lead and silver, galenite or products of oxidation were produced by processing cerussite (Conophagos, 1980; Gale and Stos-Gale, 1981). Mineral deposits containing plenty of secondary copper in Lavrion provided a substantial amount of copper production for the first 3,000 years (Gale et al., 2007). The discovery of higher-grade, third-contact mineralization in 483 BC around the Kamariza region became a milestone in the history of Lavrion mining. Thanks to this discovery, more constant and large-scaled mining and mineral processing activities could be conducted (Ross et al., 2021).

Along with Plaka and Sounion, the Megali Cariera geosite is a part of the Kamariza mining area that is accepted as the most important mining center. Thanks to its easy access and short distance from Lavrion, this mine, one of the best-known mines of Lavreotiki and an active one in both the ancient and modern ages, attracts a great number of visitors. Main ores in this area are sulfide, oxidized, and carbonate, involving lead, zinc, and iron ores mixed with main ores, silver, copper, and other elements. The mineralization shape is usually layered and irregular and in principle is within marble and the contact that marble has with schists. In research on at least three mineralization contacts, it was observed that the ore essentially intensifies on contact between marble and schist. At this geosite, there are also adits from recent mining activities as well as old adits. Unfortunately, due to destruction, not all the underground constructions can be visited today. In the open area of the mine, there are still material extraction piles/hills of small sizes in various mineral occurrences, where finding and collecting ore samples is possible. This mine was used to extract minerals from the mineralization contact in the middle. It is estimated that from here, a total of 1,500,500 tons of ore containing lead at 5–6% grade, zinc at 8%, and 70–80 gr/tons silver were extracted (Janikian and Theodossiou, 2009).

There are two main types of mineralization in the lower layer of Lavreotiki: the mineralization of sulfur ores, where intensive production was carried out and basic metals such as lead, zinc, iron, and copper were mixed, and the presence of Fe-Mn ores, the availability of which was generally limited. Primary sulfide mineralization of mixed sulfurs, constituting the main mineralization of the region, is characterized by the presence of galenite (PbS), which is rich in sphalerite (ZnS) and pyrites (FeS$_2$). A series of other sulfur minerals (for example, Cu and As) and various sulfur and salt minerals are included in the ore mineral group of the region. Dominating supporter minerals are fluorite, calcite, barite, quartz, and dolomite. Sulfur mineralization is essentially a part of carbonate forming of the Para-otokton series of Lavreotiki or the Kamariza series (lower marble, upper marble, transgressive limestone). According to the prevailing view, the origin of mineralization is epigenetic and directly (hydrothermally originated) associated with Upper Miocene magmatic activity. Its formation temperature is 280°C. The substitution type belongs to the category of carbonate ore deposits containing massive sulfur Pb-Zn-Ag. Additionally, as a result of oxidation and changing places of primary mineralization elements afterward, large accumulations of secondary ores such as carbonated zinc, smithsonite (ZnCO$_3$), and lead- or cerussite (PbCO$_3$)-originated silver carbonate were formed in the larger area of Lavreotiki. Hence, it was exposed to intensive mine extraction activities. Last, in the Plaka region of Keratea, an important magnetite ore deposit type massive skarn is located. Approximate mineral amounts, types, and grades extracted through the historical development of the region are as follows (Janikian and Theodossiou, 2009).

Ancient mining and metallurgical by-products were of three types: (1) piles of low-grade ores such as sphalerite and smithsonite that could not be processed by ancient metallurgists, often left behind near adits; (2) tailings that accumulated near the dressing plant after the ancient mineral processing; and (3) slags, which are the by-products of smelting and form huge piles (Conophagos, 1980; Dermatis, 2006). While it is estimated that in the 19th century, ancient piles of slags reached ~1,500,000 tons, what was left from the tailings were a couple of million tons (Conophagos, 1980). All ancient by-products (tailing and slags) were subjected to thorough re-processing in the 19th century (Conophagos, 1980; Dermatis, 2006). In detail, these data are as follows (Janikian and Theodossiou, 2009):

A. In ancient Lavrion (600–100 BC):

1. Extracted silvery lead ore (galenite + cerusite), 13,000,000 tons of average grade containing 20% lead and 400 gr/ton silver.
2. Metal production from hydraulic-mechanical processing and metallurgy: 1,400,000 tons of lead and 3,500 tons of silver.
3. Tailings: 1,000,000 tons of low-grade ore, 9,000,000 tons of processing tailings (plynites), 50 gr/ton of silver and 1,500,000 tons of slags containing 10% lead; 10,000,000 tons of low-grade ore containing 140 gr/ton of silver and 7% of lead.

B. Modern Lavrion (1865–1976):

1. Ilarion Roux et Compagnie (1865–1873): 60,000 tons of enriched galena as a result of the hydraulic-mechanical processing of ancient slags
   and low-grade ores (850,000 tons with an average of 7% lead).
2. Greek Mining Company (1873–1917): Ancient slags and (containing
   on average 7% Pb) 4,500,000 tons run-of-mine) of enriched galena as
   a result of hydraulic-mechanical dressing of low-grade ores.
3. FMC (1873–1977) ore extraction:

   a. Lead-zinc deposits (1,800,000 tons run-of-mine ore with an average
      grade of 45% zinc and 2% lead) 810,000 tons of enriched zinc, and
      36,000 tons of metallic lead.
   b. (3,200,000 tons run-of-mine ore with an average grade of 11% lead,
      2% zinc, 140 gr/ton silver) 64,000 tons of enriched zinc, 352,000
      tons of metallic lead, and 450 tons of silver from oxidized lead.
   c. (3,300,000 tons run-of-mine ore with an average of 6.15% lead,
      7.15% zinc, and 140 g/ton silver) 236,000 tons of metallic zinc,
      203,000 tons of metallic lead, and 460 tons of silver from mixed
      sulfur deposits.

5.5.   Mineral Processing/Ruins of Metallurgy Plants in Lavrion

Mineral processing/metallurgy plants in Lavrion played a vital role in silver
production in the Classical Greek Period (Duchène et al., 2018). Ancient mineral processing plants of the flat type would mineralize the galenite extracted
in the Lavreotiki region between the years 6th and 1st centuries BC. Mineralization processes in these plants were conducted in the following way by
using water: When the ore was extracted from the adit, in the first phase of
the process, galenite would be crushed to thin particulates in mineral processing, based on the fact that galenite or cerussite ores in their mineral form are
heavier than gangue minerals; hence they would not float on the water surface.
Galenite would be ground to the pieces smaller than 1 mm by being crushed,
and then it would dissolve into heavier and lighter pieces through the water. In
other words, heavier pieces would contain concentrated galena rich in silver,
meaning that rich ore in terms of lead and silver would stay in the gaps of the
mineral processing plants and would be melted down later. Lighter pieces, on
the other hand, would contain low-grade (tailings) ores of lead/silver. Waste
was collected in different stacks. Therefore, the first step in this process was
to melt down the ore and extract the lead that was rich in silver. The remaining
slags[3] were discarded. Next, the lead was heated, and through the help of a
constant airflow in a small melting pot in a furnace, it was melted down, and
the lead was oxidized. In the end, only pure silver would remain at the bottom

of the furnace, and the litharge without economical value would be disposed of. This two-stage procedure enabled pure silver with a ~99% grade to be extracted. After the Corinthian War, Romans used more advanced techniques, including drainage procedures to widen their adits below the water levels. In this way, they could maintain their mining activities in Lavrion (Janikian and Theodossiou, 2009; Periferakis and Paresoglou, 2019).

The most impressive and common industrial ruins among the ancient mineral processing plants in Lavrion are the ore enrichment units in different variations, while most of them are rectangular and flat (Rehren et al., 2002; Tsaimou and Frangiskos, 2008). There are some certain ruins from the old mineral processing plants and slag stacks in Lavrion. There are tens of ancient mineral processing plant ruins, most of which are flat-type mineral process-ing plants, in Lavreotiki. Spiral-type processing plants are rare. On the Poun-tazeza bay beach (Panormos), there are ruins from an old mineral processing plant where galenite would have been melted down. In addition, in a large area (~1 km² wide) along with the beach, there are tailings resulting from the mineral processing process emerging from scattered pyrite grains with flotation sands (pools). This area in particular has geo-archeological value. Thanks to these ancient ruins, it was proved that the galenite melting method had been known and used in ancient Lavrion since 1500 BC (Janikian and Theodossiou, 2009; URL-6).

Athenians developed an innovative technique when they met the prob-lem of recovery of low-grade metal ores through adjacent water storage and tunnels for mineral processing and recirculating the water needed for the dressing, considering the limited rainfall in the region. This technique is a prominent and recurring feature throughout Lavreotiki. Several hundred of these adjacent water storage and canals were found scattered throughout the mine site but in close proximity to each other (Rehren et al., 2002; Tsaimou and Frangiskos, 2008). There were well-designed water tanks and coolant res-ervoirs for rainwater collection for mineral processing because it was impos-sible to supply water from streams in the mine site in Lavrion (URL-5). This water shortage necessitated the construction of a complex drainage system used to fill mineral processing sites used for ore separation before melting. This drainage system also enabled the water of ore enrichment units to be recycled and reduces the need to transport water far away (Conophagos, 1980; Kakavogianne et al., 2019). Ancient Greeks built potable water cisterns in a wider area of Lavreotiki as well as ~200 water storage basins for feeding mineral processing plants with rainwater (Janikian and Theodossiou, 2009; URL-6).

Excavations by various organizations have revealed new excavation sites, adits and shafts, dressing plants, cisterns, and furnaces. Thus, useful and im-portant information about ancient mining and metallurgy technology has been obtained. In fact, the ruins of ancient technological plants (mostly dressing plants) remain almost undamaged despite the time that has passed. They are lasting structures that were built or carved into rocks. Since the local people used wooden constructions that could not stand against time, no similar plants were found anywhere else in this region (Janikian and Theodossiou, 2009). Modern mining and metallurgy activities in Lavrion started with metal recovery from ancient slag stacks in 1864 (Conophagos, 1980; Markouli, 2019). As mentioned in Chapter 3, Lavreotiki owes its modern development to the mineralogist A. Kordelas, who revealed the ancient mine plants in the 1860 excavations. In 1863, Kordelas suggested benefiting from ancient slags to Italian J.B. Serpieri. Thanks to this suggestion, the following year Serpieri founded the Italian-German company Roux-Seripieri-Eressynet C.A. In this way, silvery lead production started with slags and low-grade mineral processing in 1865. Next, with Castellano furnaces, small-capacity dressing plants, and machine shops with new production, the first metallurgy industry was founded (Janikian and Theodossiou, 2009). Two mining companies, the Metallurgical Company of Lavrio (1873–1927) and FMC (1875–1981), essentially founded a complete metallurgy building, which conducted mineral processing and melting down of minerals containing sulfur and created blocks out of fine-grained and high-grade ores in Kamariza and Plaka-Villia. Older Castellano-type furnaces were replaced by larger furnaces. However, the melting technology was very similar to the old techniques (Markouli, 2019). The renewal history of the mineral processing plants in Lavrion can be summarized as follows.

From 1875 to 1904, the plants of the company included administrative buildings where the enterprise was organized, mechanical mineral processing plants, hydro-mechanical mineral processing, thermal mineral processing, and lead recovery furnaces. Ores extracted from mines were first evaluated in situ. Lead, zinc, and other sulfur mixture minerals were primarily prepared mechanically by mineral processing units. Next, they would be processed metallurgically. There were two distinct procedures: calamine calcination and high-temperature heating and melting of sulfide lead. The final product contained 95% lead and was recovered as a silver-containing lead. In 1905, large-scale operating activity initiated the modernization of lead metallurgy. From then on, two new different methods of calcination were applied based on the type of the ore, and old furnaces were abandoned. Lead sulfurs containing large amounts of pyrite were now fully ustulated using the Kauffmann method, while those containing less pyrite were recovered by the Huntington-Heberlein method. Two types of Brunton furnaces were built and briquetting units were placed. As of 1905, the factory was reorganized due to the establishment of steam-powered power generation units using the gas produced at the same time. In 1913, the company expanded its metallurgical activities, which further separated valuable minerals from the tailings by melting. In the late 1920s, the company faced economic problems as cost-effective ores in the region gradually declined as lead prices continued to decline. In 1930, the operating of calamine calcination furnaces was halted since it was not profitable anymore, because economically operatable ores ran short of. Mineral processing methods were modernized, and pure lead was produced for the domestic market to handle the crisis. For this, as of 1930, the flotation method was put into practice. While silver was obtained by coating with zinc to produce chemical lead, after 1936, lead sheets were started to be produced in a rolling mill. The Mining Company in Lavrion was bought by the multinational Penarroya company. The latest innovation realized in mineral processing plants was to build a building-scale smoke filter in the early 1950s. With these innovations, the company continued its mining activities until 1989 (URL-22).

5.6. Geopark Project in Lavrion and Its Preservation

The course of human influence at Lavreotiki forms a mosaic of unprotected monuments and authentic landscapes, mostly found in ruins. This mosaic is a cultural landscape of great interest and vitality, incorporating many aspects of the landscape, where geology, mining, history, archeology, and natural features can be absorbed in different ways and finally reflected together. The unavoidable question is related to the future of such an interesting but mostly time unnoticed land (Terkenli, 1996). An important part of the geopark in Lavreotiki (natural habitat with its flora and fauna) has ecological importance. The inclusion of Sounion National Park in the European Ecological Network NATURA 2000 is clear proof of this. Systematic studies for the establishment of the geopark started in 2007. In 2009, local governments and services contributed to the formation of the geopark. The Municipal Public Benefit Agency of Lavreotiki, created by the municipality of Lavrion, is an institution that applied for the integration of the Lavreotiki Geopark into the European and Global Geoparks Networks. This institution has full responsibility for the coordination of the geopark area. The Municipal Public Benefit Agency of Lavreotiki (Enveco S.A.) practices the management plan of Lavreotiki Geopark, information, and specifications collected by its data with the close collaboration of the multidisciplinary science team of IGME. This study was conducted within the scope of the project Promotion of Geosite-Geoparks, Contribution to Sustainable Development. At the same time, this project is a project supervised by IGME and financed within the framework of the 3rd European Community Support Framework with the impact of this project, Lavreotiki region has become acceptable as a geological, natural, and cultural heritage site with international importance that can be appreciated with protection, education, and geotourism objectives. LTCP strongly supports the official application for the inclusion of Lavreotiki Geopark in the European and Global Geoparks Networks because this is an area of high importance that is renowned worldwide (URL-11). However, although attempts have been made to include Lavreotiki in the European and Global Geoparks Networks (2009) and list of UNESCO World Heritage Sites (2003, 2014), they have not been successful insofar as the necessary criteria set by UNESCO have not been met (Kayafa, 2018). Currently, there are 147 UNESCO global geoparks in 41 countries. Among them, there are five UNESCO global geoparks in Greece: Lesvos Island, Psiloritis and Sitia in Crete, Chelmos-Vouraikos in the Peloponnese, and Vikos-Aoos (Zouros, 2004; Moraiti, 2015) in Epirus. These Greek geoparks bring together extraordinary geological, paleontological, and natural features. On the other hand, according to the views of some authors, when compared to Lavrion, these geoparks are less ostentatious because Lavrion is a unique area that has interesting mineralogical, geological, mining, and archeological features compared to other geoparks in the world. The geo-touristic development of mineralogical, geological, archaeological, and metallurgical areas in Lavrion can ensure the protection of geological and cultural heritage, promote educational activities, and also set an example for sustainable development. Due to its main importance for the advancement of Ancient Athens, it's expected that archaeological, geological, mineralogical works will continue in a wide range within and around ancient mines and operating sites and to ensure the protection of the geopark. The inclusion of the Lavreotiki Geopark in the European and Global Geoparks Networks will facilitate the necessary efforts to preserve the archaeological and historical heritage of the region. In addition, it will provide a strong legal basis on the matter of how to conduct future studies on the site. In this way, maintenance of research and improvement works in the long term will be ensured. Enabling the acknowledgment of such a world heritage will ensure that future generations understand how mining, which is the oldest industry of humanity, has contributed to the social, cultural, economic, and technological development of society since the start of the Stone Age (Voudouris et al., 2021). The Lavrion mines can be regarded as an important part of both Greek and European geo-heritage. If they can be protected and operated on the necessary level, they will further strengthen their duty as a valuable resource in the areas of research, education, training, and recreation (Periferakis et al., 2019).

FMC predominantly managed its mining activities through Kamariza and Plaka. The re-operating of ancient tailings stacks, the reopening and expansion of old adits, and the destruction of many old production adits in hopes of finding high-grade materials resulted in not documenting the production and the loss of much evidence of ancient mining technology. At the same time, demolishing the old furnaces and other plants resulted in the destruction of much evidence regarding ancient mining and metallurgy (Conophagos, 1980; Photos-Jones and Ellis Jones, 1994). Today, Lavrion is a threatened industrial heritage area. Ancient archaeological excavation sites have been slowly disappearing with erosion. In the upcoming years, if no protection is provided, the adverse outcomes will increase more. Some building and machine parts left from the Industrial Age also need protection. The famous archives of the FMC need to be preserved and classified. Today, several mine shafts are filled with tons of garbage. This creates unsafe conditions for visitors. For this reason, more than ~1,000 small shafts must be located, mapped, and stabilized.

Thus, these shafts must be made safe for visitors both on the surface and underground. Restoration of ancient and modern adits will facilitate their use for touristic and scientific purposes. Despite the negative situation, the ruins of the Lavrion mining-metallurgy system are some of the best-preserved in this area of the world. In addition to the geological, mineralogical, and archaeological sites on the surface, a natural underground area that can be combined on geo-touristic routes can be used as a three-dimensional mineralogy museum. The development of various geo-touristic routes in Lavrion is seen as a top priority (Velitzelos et al., 2003; Zouros, 2004; Periferakis et al., 2019). Since there are a lot of adits that are safe in terms of structure, they can be suitably altered for visitors with relatively cheap structural improvement. The Greek state took some steps to protect and promote the rich history and culture of the region. However, more attempts are expected. Some of the ancient adits are closed and left locked. However, other than that, they are unprotected and neglected. There can also be adits that are unlocked (Figure 5.19). What is concerning is that intruders enter adits and take rare and valuable mineral samples illegally for personal gain (Periferakis et al., 2019). Unfortunately, the extreme beauty, rarity, and variety led to comprehensive commercial mineral exploitation of large underground areas by domestic and foreign merchants without any kind of control. As a result, while mineral formations were abundant in the insides of the adits, a serious decrease has emerged in time. It is necessary to protect the still-unexplored mineral pits and existing mineral occurrences. Some of the new finds should remain in situ for local observation, while most of the specimens should be exhibited in a new mineralogy museum in Lavrion (Voudouris et al., 2021). Detailed underground geological mapping could help discover new potential mineral formation sites. In addition, both on the surface and underground, it is suggested that a detailed registration and mapping of the ore areas be prepared. Using the latest developments in mineral exploration geophysics, particularly gravimeters, ancient adits can be mapped precisely. As mentioned, despite some steps taken over the years, extensive geo-protection studies are still needed in the Lavrion region (Periferakis et al., 2019).

5.7. Lavrion Minerology and Archaeology Museum

What the ruins spread over the region represent cannot be clarified without basic knowledge of their historical connections. The only place this information can be obtained is mineralogy, petrography, and archaeology museums (Kayafa, 2018). Minerals from micro- to mega-crystalline sizes are among the most spectacular collectibles displayed in museums and private collections around the world. Mineral types with extraordinary aesthetic and scientific value in Greece are in located a few areas. However, they are essentially located in three ancient mine regions: the Lavrion mines in Attica, Serifos Island in the Aegean, and the Halkidiki mines in northern Greece (Katerinopoulos and Zissimopoulou, 1994; Rieck and Rieck, 1999; Voudouris et al., 2017, 2019a; Ottens and Voudouris, 2018; Klemme et al., 2018). Here are all the historical museums in Greece, including the Mineralogy and Archeology Museum of Lavrion (URL-4):

- LMMM
- Goulandris Natural History Museum, Athens
- Museum of Mineralogy and Petrology, University of Athens
- Zoological Museum of the University of Athens
- Natural History Museum of Crete
- Rhodes Aquarium, Rhodes
- Museum of Mineralogy & Paleontology Stamatiadis, Ialysos, Rhodes
- Natural History Museum of the Lesvos Petrified Forest, Lesvos

The importance of the Lavrion region in terms of mineralogy amazes even non-geologists in the mineralogy museums in the region. The impressive variety of minerals is exhibited at IGME and various museums around the world. A part of the Lavrion mineral collection is exhibited in the Museum of Mineralogy-Petrology of the University of Athens (Figure 5.24). Thanks to this, the history of valuable metal mining, gold and silver, in particular, can be observed at the mineralogy museum and the processed metal exhibition (URL-11). This museum has a wide range of Lavrion mineral collections.

Lavrion represents a unique natural laboratory for geological, mineralogical, and geochemical research. With all its features, the Lavreotiki region can also be considered a geological museum (Katerinopoulos, 2010; Voudouris et al., 2021). Over the years, it has attracted the attention of many geological institutions and universities. This has resulted in special studies and scientific discoveries. Some minerals, such as laurionite, paralaurionite, and thorikosite have been discovered in the area. Apart from these, other minerals have recently been discovered and cannot be found anywhere else in the world, such as voudourisite (Rieck et al., 2019), nealite, georgiadesite, hilarionite, and fiedlerite.

These are crystallized components as a direct result of the melting procedure, arising from the chemical reactions between slag minerals and seawater. Lavreotiki, with its mineral features, is not only an ideal place for mineralogical research but also for geochemical and artificial crystallization studies (Periferakis et al., 2019). The Lavrion deposit contains five genetically related but distinct types of mineralization. This feature has not been observed in any other ore deposit and contains the highest number of different elements of any known mining region. Because about 15% of all minerals known for their variety and beauty and renowned worldwide are in Lavreotiki, nineral and slag samples from Lavrion are exhibited in two mineralogy museums in Kamariza in the region, which is home to hundreds of different minerals (Katerinopoulos, 2010; Voudouris et al., 2021).

When entering the entrance hall of the Lavrion Mineralogy and Archeology Museum, exhibits focusing on mining and metallurgical activities that were carried out in ancient times are encountered. In Exhibition Hall 1, there is a graphical representation of ancient mineral production and processing techniques, as well as inscriptions and metal objects related to mining operations found in metallurgical plants. In Exhibition Hall 2, stoneware vessels and tools and jewels take visitors deep into time, dating back to the Neolithic Age. Six hundred different types of original minerals found in Lavrion are unique, since some of them are not found anywhere else in the world. Rust-covered transports/wagons used by FMC mine workers for more than a century and abandoned today, colors created by oxidation on the walls, mysterious diversity, and the beauty of stalagmite formations are vividly illustrated here. These show valuable mineral treasures, the passion to produce and recover them, and the passion to discover and learn more about the underground richness and mining history not only for Greece but also all future generations of the world (URL-15).

Notes

1 This tour plan is given as a sample to understand the position of the areas and to see their proximity to each other.
2 The name "Paron" comes from this practice ("Paron" in Greek means someone exists) (Janikian and Theodossiou, 2009).
3 Slags are the waste materials produced as a result of the smelting of the ore.

6 Conclusion and Suggestions

The Lavrion mines, when considered with Greek history, have been one of the most important factors for the development of Greek culture and civilization, the domination and empowerment of Greece regarding its surroundings as an empire for thousands of years in ancient ages, and for the development of the industry of the Greek state in the modern age through the century it was built. In this process, changes occurred in the mining and metallurgy technologies of Lavrion. These changes contributed to the shaping of the industrial heritage of Lavrion in different periods. However, in the preservation of this heritage to a great extent and passing it down to future generations, LTCP project activities under the leadership and management of NTUA played a big role. Especially with financial support from the Greek state and EU and the embrace of this project by the local people, these activities could reach their goals.

Thanks to including the abandoned mine site in Lavrion, a close town to the capital of Greece (Athens), on the Tentative List of UNESCO World Heritage Sites and preserving it within the body of LTCP, this region has continued to develop in terms of finance and economics despite the economic crisis in Greece. The LTCP area has become an industrial heritage area that is an example to the world. LTCP not only generates income with its project activities but also ensures that appreciation for mining in Greece is instilled and this awareness spreads throughout the country. Across the world, there are a lot of abandoned mines/industrial areas/buildings that can be declared potential industrial heritage/geoheritage areas with similar goals. By restoring these abandoned buildings and rehabilitating the sites they are on, they can become important examples of the world's industrial/geoheritage. Building and operating an industrial, technological, and cultural park and geopark similar to Lavrion can also be achieved in any country. For this, the details stated in this study can be a helpful guide for the projects to be prepared. Thus, a park to be established in a country similar to LTCP will be able to support developments in scientific research, education, training, and recreation in different branches of science and sustainable development studies through mining and industry.

DOI: 10.4324/9781003408437-6

Project studies should be conducted to determine places that are more suitable for their declarations as industrial heritage sites in countries. For their coordination and the organization of their plans, the ministries of energy and natural resources and culture and tourism should jointly establish a commission. These studies can be organized by the university administrations closest to these regions as much as possible, including the required fields of study. By considering the conducted studies and projects, the most suitable places/sites should be taken under protection by the Ministry of Culture and Tourism. In cities where mining activities were intensely conducted, there should be studies conducted, especially regarding establishing mining/mineral museums. Locational studies on matters of archaeology, urban conservation, and mining history should be simultaneously conducted and financed. An industrial heritage application should be made to UNESCO for places where the projects have been conducted. Similarly, applications should also be made for geoparks. In this way, these areas declared industrial heritage/geoheritage sites can both provide income for the country's tourism and increase the appreciation and awareness of history/mining/earth sciences, especially among citizens in the country. Especially since the production of valuable or strategic/critical raw minerals can affect sale prices in some countries, from time to time, transnational companies try to prevent mineral extraction in these countries so as not to lower the profits of their own companies. They do this by funding unrealistic environmental movements and triggering a public uprising. Such situations can be avoided with a full mining campaign across a country, public awareness, and common-sense approaches. Industrial and mining heritage/ geoheritage sites may contribute to this issue in the future.

Considering the increase of the tourism potential of the industrial heritage site in Lavrion which is close to the capital of Greece thanks to the activities of LTCP, abandoned mines/industrial plants/sites that can also be industrial and mining heritage sites should not only be considered from the perspective of education, training, and urban conservation but also as tourism centers within their overall development, and projects should be developed accordingly.

The scientific, education, and training contributions LTCP gives to geology, mining history, general history, archaeology, and the multidisciplinary components of these sciences are remarkable. It is seen that training activities are conducted in abandoned mine sites or at the museums of their plants and that conferences, seminars, and educational activities are conducted most effectively in the historical buildings. Additionally, the fact that abandoned mine sites have become natural and protected areas where movies can be shot has increased the recognition of the region worldwide. All these features can turn an industrial heritage site into a center of attraction – as in the case of LTCP. Given that designing industrial and mining heritage/geoheritage areas in countries worldwide similar to LTCP will have great benefits for the promotion and tourism of a country, it should not escape attention that a budget to be spent on the restoration, recreation, and rehabilitation within this scope will return to these countries hundredfold. An industrial and mining heritage site or geopark area in a country can only give maximum benefit to its country if it can reach a level of development similar to LTCP. Therefore, the projects to be realized should be planned simultaneously and for the long term, and steps should be taken in this direction.

References

Abad, C.J.P., 2017. The post-industrial landscapes of Riotinto and Almadén, Spain: Scenic value, heritage and sustainable tourism. *Journal of Heritage Tourism*, 12 (4), 331–346. https://doi.org/10.1080/1743873X.2016.1187149

Agricola, G., 1950. *De Re Metallica 1556*, Translated from the Latin first edition into English. Hoover, H.C. ve Hoover, L.H., Dover Publications Inc.

Ahmad, S., Jones, D., 2013. Investigating the mining heritage significance for Kinta District, the industrial heritage legacy of Malaysia. *Procedia – Social and Behavioral Sciences*, 105, 445–457. https://doi.org/10.1016/j.sbspro.2013.11.047

Albrecht, H., 2013. What does the industrial revolution signify? Chapter 2 in book. *Industrial Heritage Re-tooled: The TICCIH Guide to Industrial Heritage Conservation*. J. Douet (editor), Routledge, Taylor & Francis Group, 1st edition. https://doi.org/10.4324/9781315426532

Alfrey, J., Putnam, T., 1992. *The Industrial Heritage: Managing Resources and Uses*, Routledge, Taylor & Francis Group, 1st edition. https://doi.org/10.4324/9780203392911

Aliağaoğlu, A., 2004. Sosyo-kültürel miras turizmi ve Türkiye'den örnekler (Sociocultural heritage tourism and examples from Turkey). *Journal of Geographical Sciences*, 2 (2), 50–64. https://doi.org/10.1501/Cogbil_0000000045

Alpan, A., 2012. Eski sanayi alanlarının yazındaki yerine ve endüstri arkeolojisinin tarihçesine kısa bir bakış (A brief look at the place of old industrial areas in the literature and the history of industrial archeology). *Planlama (Planning)*, 1–2, 21–28. Available via: <www.researchgate.net/publication/346658900_Eski_Sanayi_Alanlarinin_Yazindaki_Yerine_ve_Endustri_Arkeolojisinin_Tarihcesine_Kisa_Bir_Bakis>

Álvarez, J.P.F., Pieiga, A.G., Lázare, C.J.S., 2010. New concepts in reassessing mining heritage: A study and its implications from the Ancient Iron Mine of Llumeres (North Spain). *Journal of Cultural Heritage*, 11 (2), 172–179. https://doi.org/10.1016/j.culher.2010.01.001

Apostolopoulou-Kakavoyanni, O., 2008. Βίος και πολιτισμός στη Λαυρεωτική κατά τους προϊστορικούς χρόνους. Ιn: Πρακτικά Θ' Επιστημονικής Συνάντησης ΝΑ Αττικής, Λαύρειο Αττικής 13–16 Απριλίου 2000, 37–51. Quoted from (Kayafa, 2018).

Ateş, Y., 2016. The significance of historical mining sites as cultural/heritage resources- A case study of Zilan Historical Mining Site, Erçiş, Van, Turkey. *MT Bilimsel Journal of Underground Resources*, 5 (9), 15–24. Available via: <https://dergipark.org.tr/tr/download/article-file/366506>

94   References

Aydın, E.Ö., 2013. *"Korumanın yeni dinamikleri", restorasyon ve koruma ilkeleri ("New dynamics of conservation", restoration and conservation principles).* B. Y. Uçkan (editor), Open Education Faculty Publication, No: 1813, Eskişehir.

Ayhan, A., 2009. A mining history inscribed in culture: Kopalnia Soli. *Mining Turkey Journal*, 1, 12–18. Available via: <www.mtmagaza.com/wp-content/uploads/2018/05/Madencilik-Turkiye-Dergisi-Sayi-1-v3g546nm.pdf>

Ballesteros, E.R., Ramírez, M.H., 2007. Identity and community – Reflections on the development of mining heritage tourism in Southern Spain. *Tourism Management*, 28 (3), 677–687. https://doi.org/10.1016/j.tourman.2006.03.001

Bandarin, F., Oers, R.O., 2012. *The Historic Urban Landscape Managing Heritage in an Urban Century*, A John Wiley & Sons Publication. Quoted from (Karadağ and İncedere, 2020).

Barbosa, S., Amaro, S., 2020. *D3.2 Transitional Landscape Profiles in RIS Countries and Factors of Influence*, Revitalising Post Mining Regions (ReviRis), EIT Raw Materials.

Baykan, D., 2015. Antik metalurji uygulamaları (Ancient metallurgy procedure). *MT Bilimsel, Journal of Underground Resources*, 5 (9), 61–67. Available via: <https://dergipark.org.tr/en/pub/mtb/issue/32055/354907>

Belavilas N., 2012. Οι πρώτοι οικισμοί εργατών στην Ελλάδα. Λαύριο, Δραπετσώνα, Ελευσίνα. Paper based on an announcement for the Conference: Εργάζομαι άρα κατοικώ: η περίπτωση του συγκροτήματος κατοικιών των μεταλλείων Μπάρλου στο Δίστομο Βοιωτίας, των Δ. & Σ. Αντωνακάκη, University of Patras, Department of Architecture, Patra, March 14, pp. 1–15. Accessed date: 21/014/2015. Available via: <http://courses.arch.ntua.gr/112052.html>, Quoted from (Kayafa, 2018).

Belavilas, N., Papastefanaki, L., 2009. Εισαγωγή: τα ορυχεία του Αιγαίου ως βιομηχανικά μνημεία. *Ορυχεία στο Αιγαίο, Βιομηχανική Αρχαιολογία στην Ελλάδα*, N. Belavilas and L. Papastefanaki (editors), Melissa, pp. 18–23. Quoted from (Kayafa, 2018).

Benardos, A.G., Kaliampakos, D.C., 2006. Design of an underground hazardous waste repository in Greece. *Tunnelling and Underground Space Technology*, 21, 3–4. https://doi.org/10.1016/j.tust.2005.12.053

Benardos, A., Kaliampakos, D., 2013. Underground solutions for urban waste management: Status and perspectives. *International Solid Waste Association (ISWA)*, Available via: <www.researchgate.net /publication/318503365_Underground_Solutions_for_Urban_Waste_Management_Status_and_Perspectives>

Berger, S., Golombek, J., Wicke, C., 2018. A post-industrial mindscape? The mainstreaming and touristification of industrial heritage in the Ruhr Chapter 4 in book. *Industrial Heritage and Regional Identities*, C. Wicke, S. Berger and J. Golombek (editors), Routledge, Taylor & Francis Group, 1st edition. https://doi.org/10.4324/9781315281179

Berger, S., Pickering, P., 2018. Regions of heavy industry and their heritage – between identity politics and 'touristification' Chapter 10 in book. *Industrial Heritage and Regional Identities*, C. Wicke, S. Berger and J. Golombek (editors), Routledge, Taylor & Francis Group. https://doi.org/10.4324/9781315281179

Bergeron, L., 2013. e heritage of the industrial society Chapter 4 in book. *Industrial Heritage Re-Tooled: The TICCIH Guide to Industrial Heritage Conservation*, J. Douet (editor), Routledge, Taylor & Francis Group, 1st edition. https://doi.org/10.4324/9781315426532

Berkenbosch, K., Groote, P., Stoffelen, A., 2022. Industrial heritage in tourism marketing: Legitimizing post-industrial development strategies of the Ruhr Region, Germany. *Journal of Heritage Tourism*, 17 (3), 327–341. https://doi.org/10.1080/1743
873X.2022.2026364

Bilgiç, M., 2013. The world's first mining ecomuseum: Ecomuseo Delle Miniere E
Della Val Germanasca-Prali. *Mining Bulletin*, 105, 53. Available via: <www.maden.
org.tr/resimler/ekler/edfa39168e4778c_ek.pdf>

Bozzuto, P., Geroldi, C., 2021. The former mining area of Santa Barbara in Tuscany and
a spatial strategy for its regeneration. *The Extractive Industries and Society*, 8 (1),
147–158. https://doi.org/10.1016/j.exis.2020.09.007

Brilha, J., 2021. Geoheritage. *Encyclopedia of Geology*, 2nd edition, pp. 569–578.
https://doi.org/10.1016/B978-0-12-409548-9.12106-2

Büyükarslan, B., Güney, E.D., 2013. Endüstriyel miras yapılarının yeniden
işlevlendirilme süreci ve İstanbul Tuz Ambarı örneği (Re-use process of industrial
heritage building case: "Tuz Ambarı" in Istanbul). *Beykent University Journal of Science and Engineering*, 6 (2), 31–58. Available via: <https://dergipark.org.tr/tr/pub/
bujse/issue/3849/51492>

Büyüktaşkın, H.A., Türkel, E., 2019. Gasometer structures from past to present and
possible re-functioning proposals for Dolmabahçe Gasometer. *Turkish Academy
of Sciences Journal of Cultural Inventory*, 19 (1), 55–74. https://doi.org/10.22520/
tubaked.2019.19.004

Casanelles, E., 2013. TICCIH's charter for industrial heritage Chapter 33 in book. *Industrial Heritage Re-Tooled: The TICCIH Guide to Industrial Heritage Conservation*, J. Douet (editor), Routledge, Taylor & Francis Group, 1st edition. https://doi.
org/10.4324/9781315426532

Casanelles, E., Douet, J., 2013. Conserving industrial artefacts Chapter 27 in book.
*Industrial Heritage Re-Tooled: The TICCIH Guide to Industrial Heritage Conserva-
tion*, J. Douet (editor), Routledge, Taylor & Francis Group, 1st edition. https://doi.
org/10.4324/9781315426532

Cassel, S.H., Pashkevich, A., 2011. Heritage tourism and inherited institutional structures: The case of Falun Great Copper Mountain. *Scandinavian Journal of Hospitality and Tourism*, 11 (1), 54–75. https://doi.org/10.1080/15022250.2011.540795

Çetinkaya, K., 2020. A mine meeting with tourism will be more recognizable. *Mining
Turkey Magazine*, 91, 84–88. Available via: <https://madencilikturkiye.com/wp-
content/uploads/2018/09/Madencilik-Turkiye-Dergisi-Sayi-91-jf659jg4d7943.
pdf>

Chadoumelis, A., 2015. Strategic analysis and planning (2015–2020): Updated version
February 2016. *Lavrion: NTUA, Lavrion Technological and Cultural Park*. Quoted
from (Chatzi Rodopoulou, 2020).

Chamber of Mining Engineers, 2011. Report on the establishment of a mining museum in Zonguldak, a mining city, and in our cities where mining activities are
intense, September 26. Available via: <www.maden.org.tr/genel/ bizden_detay.
php?kod=6842&tipi=5&sube=0>

Chamber of Mining Engineers, 2014. Mineral museum. *Mining Bulletin*, 109, 17. Available via: <www.maden.org.tr/resimler/ekler/2392f623db5aa3a_ek.pdf>

Chatzi Rodopoulou, T., 2020. Control Shift: European industrial heritage reuse in review. Volume 1 and 2. *A+BE Architecture and the Built Environment*. https://doi.
org/10.7480/abe.2020.13

Che, D., 2015. Developing a heritage tourism attraction in a working salt mine: The Kansas underground salt museum Chapter 8 in book. *Mining Heritage and Tourism: A Global Synthesis*, M.V. Conlin and L. Jolliffe (editors), Routledge, Taylor & Francis Group, 1st edition, https://doi.org/10.4324/9780203865507

Cho, H., 2014. Advocacy coalition for historic preservation in the U.S.: Changes in motivations. *The Journal of Arts Management, Law, and Society*, 44 (4), 234–245. https://doi.org/10.1080/10632921.2014.960628

Çiftçi, Y., Güngör, Y., 2016. Proposals for the standard presentation of elements of natural and cultural heritage within the scope of geopark projects. *Bulletin of the Mineral Research and Exploration*, 153, 223–238. https://doi.org/10.19111/bmre.80846

Cigna, F., Tapete, D., Lee, K., 2018. Geological hazards in the UNESCO World Heritage sites of the UK: From the global to the local scale perspective. *Earth-Science Reviews*, 176, 166–194. https://doi.org/10.1016/j.earscirev.2017.09.016

Çiner, B., 2021. Peyzaja dair önemli kavramlar nelerdir? (What are the important concepts about landscape?) Available via: <www.evimveailem.com/peyzaja-dair-onemli-kavramlar-nelerdir/>

Cole, D., 2004. Exploring the sustainability of mining heritage tourism. *Journal of Sustainable Tourism*, 12 (6), 480–494. https://doi.org/10.1080/09669580408667250

Cole, D., 2008. Museum marketing as a tool for survival and creativity: the mining museum perspective. *Museum Management and Curatorship*, 23 (2), 177–192. https://doi.org/10.1080/09647770701865576

Conesa, H.M., Schulin, R., Nowack, B., 2008. Mining landscape: A cultural tourist opportunity or an environmental problem?: The study case of the Cartagena – La Unión Mining District (SE Spain) Ecological Economics, 64 (4), 690–700. https://doi.org/10.1016/j.ecolecon.2007.06.023

Conlin, M., Jolliffe, L., 2015. What happens when mining leaves? Chapter in book. *Mining Heritage and Tourism: A Global Synthesis*, M.V. Conlin and L. Jolliffe (editors), Routledge, Taylor & Francis Group, 1st edition, pp. 3–10. https://doi.org/10.4324/9780203865507|

Conophagos, C., 1980. *Το αρχαίο Λαύριο και η ελληνική τεχνική παραγωγής του αργύρου*. Quoted from (Kayafa, 2018) and (Voudouris, 2021).

Cossons, N., 2013. Why preserve the industrial heritage? Chapter 1 in book. *Industrial Heritage Re-Tooled: The TICCIH Guide to Industrial Heritage Conservation*, J. Douet (editors), Routledge, Taylor & Francis Group, 1st edition. https://doi.org/10.4324/9781315426532

Crofts, R., Gordon, E.J., 2014. Geoconservation in protected areas. *Parks*, 20 (2), 61–76. https://doi.org/10.2305/IUCN.CH.2014.PARKS-20–2.RC.en

Ćurčić, N., Garača, V., Vukosav, S., Bradić, M., 2015. Regeneration of industrial heritage in terms of sustainable tourism development. International Scientific Conference Geobalcanica, 5–7 June, pp. 471–479.

Damigos, D., Kaliampakos, D., 2012. Emerging value of brownfields regeneration. *International Journal of Sustainable Development and Planning*, 7 (2), 173–185. https://doi.org/10.2495/SDP-V7-N2–173–185

David, A., Kideckel, D.A., 2018. A post-industrial mindscape? The mainstreaming and touristification of industrial heritage in the Ruhr. Chapter 6 in book. *Industrial Heritage and Regional Identities*, C. Wicke, S. Berger and J. Golombek (editors), Routledge, Taylor & Francis Group, 1st edition. https://doi.org/10.4324/9781315281179

Del pozo, P.B., González, P.A., 2012. Industrial heritage and place identity in Spain: From monuments to landscapes. *Geographical Review*, 102 (4), 446–464. https://doi.org/10.1111/j.1931-0846.2012.00169.x

Dermatis, G., 1994. *Scenery and Monuments of Laureotiki*, Publication of the Municipality of Laureotiki. Quoted from (Periferakis et al., 2019).

Dermatis, G, 2017. *ΛΑΥΡΙΟ Η ΒΟΗ ΤΟΥ ΧΡΟΝΟΥ, Το νεότερο Λαύριοτης Μεταλλείας και της Μεταλλουργίας-LAVRIO*, THE RUMBLE OF TIME,The Modern Lavrio of Mining and Metallurgy, ΕΜΕΛ (Research Associationon Lavreotiki).

Dermatis, G., Marmani, M., Belavilas, N., Chatzi-Rodopoulou, T., 2010. A twenty Year Debt in history: Mining and metallurgy Museum of Lavrion. *TICCIH GR Journal*, C.1, 88–91. Quoted from (Chatzi Rodopoulou, 2020).

Dermatis, G.N., 2006. *Ως εφάνη τα μέταλλα τα εν Μαρωνεία – Τα μεταλλεία της Καμάριζας στο Λαύρειο*, Kamariza. Quoted from (Kayafa, 2018).

Dewar, K., Miller, R.F., 2015. Geotourism, mining and tourism development in the Bay of Fundy, Canada. Chapter 18 in book. *Mining Heritage and Tourism: A Global Synthesis*, M.V. Conlin and L. Jolliffe (editors), Routledge, Taylor & Francis Group, 1st edition. https://doi.org/10.4324/9780203865507|

Dieterich-Ward, A., 2018. Industrial heritage and regional identity in metropolitan Pittsburgh. Industrial heritage in Newcastle and the Hunter Valley, Australia. Chapter 9 in book. *Industrial Heritage and Regional Identities*, C. Wicke, S. Berger, and J. Golombek (editors), Routledge, Taylor & Francis Group, 1st edition. https://doi.org/10.4324/9781315281179

Douet, J., 2013. *Industrial heritage re-tooled: The TICCIH guide to industrial heritage conservation*, Routledge, Taylor & Francis Group. https://doi.org/10.4324/9781315426532

Doukellis, P.N., 2009. Αναζητώντας το τοπίο. *Το ελληνικό τοπίο – Μελέτες ιστορικής Γεωγραφίας και πρόσληψης του τόπου*, P.N. Doukellis (editor), Estia Publications, 3rd edition, pp. 13–31. Quoted from (Kayafa, 2018).

Duchène, S., van Liefferinge, K., Kayafa, M., 2018. The ore-processing workshops. *Exploring Thorikos*, F. Roald and Maud (editors), The Section of Mediterranean Archaeology, Department of Archaeology, Ghent University, pp. 43–45. Quoted from (Voudouris et al., 2021).

Ebert, W., 2013. Industrial heritage tourism. Chapter 28 in book. *Industrial Heritage Re-Tooled: The TICCIH Guide to Industrial Heritage Conservation*, J. Douet (editor), Routledge, Taylor & Francis Group, 1st edition. https://doi.org/10.4324/9781315426532

Economopoulos, J., 1996. Mining activities in ancient Greece from the 7th to the 1st centuries BC. *Mining History Journal*, 109–114. Available via: <www.mininghistoryassociation.org/Journal/MHJ-v3-1996-Economopoulos.pdf> Quoted from (Periferakis et al., 2019).

Edensor, T., 2005. *Industrial Ruins: Aesthetics, Materiality and Memory*. Berg. Quoted from (Karadağ and İncedere, 2020).

Edwards, J.A., Coit, C.J.L., 1996. Mines and quarries: Industrial heritage tourism. *Annals of Tourism Research*, 23 (2), 341–363. https://doi.org/10.1016/0160-7383(95)00067-4

Efesiou, I., 2011. Lavrion technological cultural park: 1992–2010, survey, documentation, constructional analysis and assessment of the building complex Chapter 7 in book. *Chronocity: Sensitive Interventions in Historic Environment*, D. Babalis (editor), Alinea Editrice, pp. 30–34.

Ehsani, A., Yazıcı, E.Y., 2018. Anadolu'da bakır madenciliği ve kullanımının kısa tarihçesi. (A brief history of copper mining and use in Anatolia). *MT Bilimsel: Journal of Underground Resources*, 5 (9), 43–48. Available via: <https://dergipark.org.tr/en/download/article-file/366509>

Eklund, E., 2018. Industrial heritage in Newcastle and the Hunter Valley, Australia. Chapter 8 in book. *Industrial Heritage and Regional Identities*, C. Wicke, S. Berger and J. Golombek (editors), Routledge, Taylor & Francis Group, 1st edition. https://doi.org/10.4324/9781315281179

Elhan, S., 2009. Evaluating the industrial heritage in the context of urban memory: Case study Istanbul Halic. Master thesis. Istanbul Technical University Institute of Science and Technology. Quoted from (Kaya and Yılmaz, 2018).

Ergünalp, D., 2020. Georgius Agricola, (alias Georg Bauer). *Turkish Mining Development Foundation (YMGV), Sector Mining Journal*, 77, 42–43. Available via: <https://cdn.ymgv.org.tr/cdn/uploads/dosyalar/sektormaden-aralik2020.pdf>

Falconer, K., 2013. Legal protection. Chapter 12 in book. *Industrial Heritage Re-Tooled: The TICCIH Guide to Industrial Heritage Conservation*, J. Douet (editor), Routledge, Taylor & Francis Group, 1st edition. https://doi.org/10.4324/9781315426532

Ferraro, F.X., Schilling, M.E., Baeza, S., Oms, O., Sá, A.A., 2020. Bottom-up strategy for the use of geological heritage by local communities: Approach in the "Litoral del Biobío" Mining Geopark project (Chile). *Proceedings of the Geologists' Association*, 131 (5), 500–510. https://doi.org/10.1016/j.pgeola.2020.06.001

Fidan, E., 2015. Ore deposits of anatolia used in prehistorical times. *MT Bilimsel, Journal of Underground Resources*, 5 (9), 49–59. Available via: <https://dergipark.org.tr/en/pub/mtb/issue/32055/354906>

Florentina-Cristina, M., Andreea-Loreta, C., Daniel, P., 2015. Roşia Montană, Romania: Industrial heritage in situ, between preservation, controversy and cultural recognition. *Industrial Archaeology Review*, 37 (1), 5–19. https://doi.org/10.1179/0309 072815Z.00000000039

Florentina-Cristina, M., George-Laurențiu, M., Andreea-Loreta, C., Constantin, D.C., 2014. Conversion of industrial heritage as a vector of cultural regeneration. *Procedia – Social and Behavioral Sciences*, 122, 162–166. https://doi.org/10.1016/j.sbspro.2014.01.1320

Fontaine, M., 2018. Regional identity and industrial heritage in the mining area of Nord-Pas-de-Calais. Chapter 3 in book. *Industrial Heritage and Regional Identities*, C. Wicke, S. Berger and J. Golombek (editors), Routledge, Taylor & Francis Group, 1st edition. https://doi.org/10.4324/9781315281179

Forget, M., Rossi, M., 2021. Mining region value and vulnerabilities: Evolutions over the mine life cycle. *The Extractive Industries and Society*, 8 (1), 176–187. https://doi.org/10.1016/j.exis.2020.07.010

Fotiou, T., Monemvasitou, A., Kafritsa, M., Papaioannou, A., 2011. Lavrion technological cultural park: 1992–2010, Reusing the past in order to design the present. Chapter 6 in book. *Chronocity: Sensitive Interventions in Historic Environment*, pp. 27–29.

Fragner, B., 2013. Adaptive re-use. Chapter 14 in book. *Industrial Heritage Re-Tooled: The TICCIH Guide to Industrial Heritage Conservation*, J. Douet (editor), Routledge, Taylor & Francis Group, 1st edition. https://doi.org/10.4324/9781315426532

Frew, E., 2015. Transforming working mines into tourist attractions: Conceptual and practical considerations. Chapter 7 in book: *Mining Heritage and Tourism: A Global*

Synthesis, M.V. Conlin and L. Jolliffe (editors), Routledge, Taylor & Francis Group, 1st edition, pp. 72–83. https://doi.org/10.4324/9780203865507|

Gałaś, A., Gałaś, S., 2011. Landscape: Functional valorisation and sustainable development in the Valley of the Volcanoes in Peru. *Polish Journal of Environmental Studies*, 20 (4A), 67–71.

Gałaś, A., Gałaś, S., Zavala, B., Churata, D., 2017. Chances of geotourism development in the colca and the volcanoes of Andagua Geopark (Peru), 17th International Multidisciplinary Scientific GeoConference, Ecology, Economics, Education and Legislation, 29th June–5th July, Volume: 17, pp. 197–204.

Gałaś, A., Haghighat-Khah, R.E., Cuber, P., Benavente, M., Gorfinkiel, D., Gałaś, S., 2022. The impact of COVID-19 pandemic on halting sustainable development in the Colca y Volcanes de Andagua UNESCO Global Geopark in Peru – Prospects and Future. *Sustainability*, 14, 4043. https://doi.org/10.3390/su14074043

Gałaś, A., Paulo, A., Gaidzik, K., Zavala, B., Kalicki, T., Churata, D., Gałaś, S., Mariño, J., 2018a. Geosites and geotouristic attractions proposed for the project geopark Colca and Volcanoes of Andagua, Peru. *Geoheritage*, 10, 707–729. https://doi.org/10.1007/s12371-018-0307-y

Gałaś, A., Tyszer, M., Gałaś, S., 2018b. Using GIS to predict potential environmental conflicts in the Colca and Andagua Volcanoes Geopark (Peru). "Public Recreation and Landscape Protection – With Nature Hand in Hand!" Conference Proceeding, 2–4 May, pp. 424–428.

Gałaś, S., Gałaś, A., Benavente, M., Tyszer, M., 2018c. Forecast of environmental impact of tourism development in the Geopark Colca and Andagua Volcanoes in the southern Peru. "Public Recreation and Landscape Protection – with nature hand in hand!" Conference Proceeding, 2–4 May, pp. 186–191.

Gale, N.H., Kayafa, M., Stos-Gale, Z.A., 2007. Further evidence for Bronze Age production of copper from ores in the Lavrion ore district, Attica, Greece. *Proceedings of the 2nd International Conference: Archaeometallurgy in Europe*, A. Giumlia-Mair, P. Craddock, A. Hauptmann, J. Bayley and M. Cavallini (editors). Quoted from (Kayafa, 2018).

Gale, N.H., Stos-Gale, Z.A., 1981. Cycladic lead and silver metallurgy. *Annual of the British School at Athens*, 76, 169–224. Quoted from (Voudouris et al., 2021).

Gámez, B.O., 2013. Industrial archives and company records. Chapter 14 in book. *Industrial Heritage Re-Tooled: The TICCIH Guide to Industrial Heritage Conservation*, J. Douet (editor), Routledge, Taylor & Francis Group, 1st edition. https://doi.org/10.4324/9781315426532

Garcia, M.G., Nascimento, M.A.L., Mansur, K.L., Pereira, R.G.F.A., 2022. Geoconservation strategies framework in Brazil: Current status from the analysis of representative case studies. *Environmental Science & Policy*, 128, 194–207. https://doi.org/10.1016/j.envsci.2021.11.006

García-Sánchez, L., Canet, C., Cruz-Pérez, M.A., Morelos-Rodríguez, L., Salgado-Martínez, E., Corona-Chávez, P., 2021. A comparison between local sustainable development strategies based on the geoheritage of two post-mining areas of Central Mexico. *International Journal of Geoheritage and Parks*, 9 (4), 391–404. https://doi.org/10.1016/j.ijgeop.2021.10.001

Gatelier, E., Ross, D., Phillips, L., Suquet, J.B., 2022. A business model innovation methodology for implementing digital interpretation experiences in European cultural heritage attractions. *Journal of Heritage Tourism*, 17 (4), 391–408. https://doi.org/10.1080/1743873X.2022.2065920

Ghazi, J.M., Hamdollahi, M., Moazzen, M., 2021. Geotourism of mining sites in Iran: An opportunity for sustainable rural development. *International Journal of Geoheritage and Parks*, 9 (1), 129–142. https://doi.org/10.1016/j.ijgeop.2021.02.004

Gill, J.C., 2017. Geology and the sustainable development goals. *Episodes*, 40 (1), 70–76. https://doi.org/10.18814/epiiugs/2017/v40i1/017010

Gordon, J.E., Crofts, R., Gray, M., Tormey, D., 2021. Including geoconservation in the management of protected and conserved areas matters for all of nature and people. *International Journal of Geoheritage and Parks*, 9 (3), 323–334. https://doi.org/10.1016/j.ijgeop.2021.05.003

Gouthro, M.B., Palmer, C., 2015. Pilgrimage in heritage tourism: Finding meaning and identity in the industrial past. Chapter 4 in book. *Mining Heritage and Tourism: A Global Synthesis*, M.V. Conlin and L. Jolliffe (editors), Routledge, Taylor & Francis Group, 1st edition. https://doi.org/10.4324/9780203865507|

Gray, M., 2019. Geodiversity, geoheritage and geoconservation for society. *International Journal of Geoheritage and Parks*, 7 (4), 226–236. https://doi.org/10.1016/j.ijgeop.2019.11.001

Güdü Demirbulat, Ö., Karaman, S., 2014. Trabzon Ayasofya Müzesi'nin camiye dönüştürülmesine ilişkin turist rehberlerinin değerlendirmesi (Assessments of tourist guides about converting of Trabzon Hagia Sophia Museum to Mosque). *Balıkesir University The Journal of Social Sciences Institute*, 17 (32), 37–54. https://doi.org/10.31795/baunsobed.645497

Gülersoy, N.Z., 2000. Conservation problems of historic houses in the central area of Istanbul. *Housing and Urban Policies for Low-income People in the Central Areas of Istanbul and Sao Paulo*, C.N. Maritano, L. Vitale, H. Turgut, G. Erkut, N.Z. Gülersoy and A. Piccini (editors), Turin.

Gülersoy, N.Z., 2003. Social housing and urban conservation: Case Study on Istanbul Historic Peninsula-Turkey. Partnership for World Heritage Cities: Culture as a Vector for Sustainable Urban Development, UNESCO World Heritage Center, Urbino, Italy, 10–13 November, World Heritage Papers, (No. 9), pp. 54–57.

Gülersoy, N.Z., 2006. Istanbul's cultural and natural assets to be preserved. *Architectural Guide to Istanbul, Historic Peninsula, Chamber of Architects of Turkey*, A. Batur (editor), Istanbul Metropolitan Branch, pp. 23–31.

Gülersoy, N.Z., 2007. Historical peninsula and future projections. *IMIAD Istanbul 2007 Workshop, International Master of Architectural Design*, ITU, Faculty of Architecture, Department of Interior Architecture, pp. 31–38.

Gülersoy, N.Z., Özsoy, A., Tezer, A., Yiğiter Genli, R., Günay, Z., 2009. Strategic quality planning in historic urban environment. *AIZ, ITU Journal of the Faculty of Architecture*, 6 (1), 109–125. Available via: <https://jag.journalagent.com/itujfa/pdfs/ITUJFA-26213-DOSSIER_ARTICLES-ZEREN_GULERSOY.pdf>

Gülersoy, N.Z., Tezer, A., Yiğiter, R., Koramaz, K., Günay, Z., 2008a. *İstanbul Project: İstanbul Historic Peninsula Conservation Study; Volume 1: Conservation of Cultural Assets in Turkey*, Istanbul Technical University Faculty of Architecture, Ofset Yapımevi.

Gülersoy, N.Z., Tezer, A., Yiğiter, R., Koramaz, K., Günay, Z., 2008b. *İstanbul Project: İstanbul Historic Peninsula Conservation Study; Volume 2: Zeyrek Case*, Istanbul Technical University Faculty of Architecture, Ofset Yapımevi.

Gülersoy, N.Z., Tezer, A., Yiğiter, R., Koramaz, K., Günay, Z., 2008c. *İstanbul Project: İstanbul Historic Peninsula Conservation Study; Volume 3: Suleymaniye Case*, Istanbul Technical University Faculty of Architecture, Ofset Yapımevi.

Gülersoy, N.Z., Tezer, A., Yiğiter, R., Koramaz, K., Günay, Z., 2008d. *İstanbul Project: İstanbul Historic Peninsula Conservation Study; Volume 4: Yenikapi Case*, Istanbul Technical University Faculty of Architecture, Ofset Yapımevi.

Gürler, M., 2001. Anıt nitelikli jeolojik oluşumlar ve koruma çalışmaları (Monumental geological formations and conservation studies). *Mavi Gezegen, Chamber of Geological Engineers*, 4, 10–11.

Habibi, T., Ruban, D.A., 2017. Outstanding diversity of heritage features in large geological bodies: The Gachsaran Formation in southwest Iran. *Journal of African Earth Sciences*, 133, 1–6. https://doi.org/10.1016/j.jafrearsci.2017.05.010

Hansoy, P., Gülersoy, N.Z., 2013. Historical creative cluster: Babıali – Istanbul. *International Journal of Economics and Finance Studies*, 5 (2), 74–85. Available via: <https://dergipark.org.tr/tr/pub/ijefs/issue/26161/275571>

Hardesty, D.L., 1990. Evaluating site significance in historical mining districts. *Historical Archeology*, 24 (2), 42–51. https://doi.org/10.1007/BF03374128

Harsft, J., 2015. Utilizing the past: Valorizing post-mining potential in Central Europe. *The Extractive Industries and Society*, 2 (2), 217–224. https://doi.org/10.1016/j.exis.2015.01.003

Hashimoto, A., Telfer, D.J., 2017. Transformation of Gunkanjima (Battleship Island): from a coalmine island to a modern industrial heritage tourism site in Japan. *Journal of Heritage Tourism*, 12 (2), 107–124. https://doi.org/10.1080/17438 73X.2016.1151884

Hatjimichalis, K., 2010. *Greekscapes. Το σύγχρονο ελληνικό πολιτιστικό τοπίο: θεωρητικό πλαίσιο, μεθοδολογία, συνθετικές παρατηρήσεις*, John S. Latsis Foundation. Quoted from (Kayafa, 2018).

Hewison, R., 1987. *The Heritage Industry*, Methuen Publishing.

Hoa, C.N.D., Chesworth, N., Jolliffe, L., 2015. Planning for the future: Tourism options for an open pit coal mine at Ha Long Bay, Vietnam. Chapter 15 in book. *Mining Heritage and Tourism: A Global Synthesis*, M.V. Conlin and L. Jolliffe (editors), Routledge, Taylor & Francis Group, 1st edition. https://doi.org/10.4324/ 9780203865507|

Hofer, A., 2017. The Sulzer/SLM site in Winterthur, Switzerland: From the factory to the New Town – The Reinvention of the city. Chapter 6 in book. *Industrial Heritage Sites in Transformation: Clash of Discourses*, H.A. Mieg and H. Oevermann (editors), Routledge, Taylor & Francis Group, 1st edition. https://doi.org/10.4324/9781315797991

Hollick, M., 2015. Sustaining the benefits of heritage mining for site, city and region: Exploring the success of Sovereign Hill outdoor museum. Chapter 8 in book. *Mining Heritage and Tourism: A Global Synthesis*, M.V. Conlin and L. Jolliffe (editors), Routledge, Taylor & Francis Group, 1st edition. https://doi.org/10.4324/9780203865507

Hooker, R., 1996. Polis city-state. Available via: <https://web.archive.org/web/20110104052653/www.wsu.edu:8080/~dee/GLOSSARY/POLIS.HTM>

Hose, T.A., 1995. Selling the story of Britain's stone. *Environmental Interpretation*, 10 (2), 16–17.

Hospers, G.J., 2002. Industrial heritage tourism and regional restructuring in the European Union. *European Planning Studies*, 10 (3), 397–404. https://doi.org/ 10.1080/09654310220121112

Hroncek, P., Rybar, P., 2016. Relics of manual rock disintegration in historical underground spaces and their presentation in mining tourism. *Acta Montanistica Slovaca*, 21 (1), 53–66. https://doi.org/10.3390/ams21010053

102 References

Huang, G., Zhou, W., Ali, S., 2011. Spatial patterns and economic contributions of mining and tourism in biodiversity hotspots: A case study in China. *Ecological Economics*, 70 (8), 1492–1498. https://doi.org/10.1016/j.ecolecon.2011.03.010

Hughes, S., 2013. Ematic World Heritage Studies. Chapter 24 in book. *Industrial Heritage Re-Tooled: The TICCIH Guide to Industrial Heritage Conservation*, J. Douet (editor), Routledge, Taylor & Francis Group, 1st edition. https://doi.org/10.4324/9781315426532

ICMM, 2003. *Position statement – mining and protected areas. International Council on Mining and Metals (ICMM). Available via: <https://www.goldfields.com/pdf/sustainbility/sustainability-reporting/international-council-on-mining-and-metals-(icmm)/2021/annexure-2-icmm-position-statements-compliance.pdf>*

ICOMOS, 2016. The Industrial heritage, a new heritage. *International Council on Monuments and Sites (ICOMOS)*. Available via: <www.icomos.org/18thapril/2006/18april2006-4.htm>

Ifko, S., 2016. Comprehensive management of industrial heritage sites as a basis for sustainable regeneration. *Procedia Engineering*, 161, 2040–2045. https://doi.org/10.1016/j.proeng.2016.08.800

Janikian, Z., Theodossiou, I., 2009. *Lavrion and Surrounding Area. Project: Designation of Geosites – Geoparks, Contribution to Sustainable Development*, The series "Geotrails in Greece" is published by the Greek Institute of Geology and Mineral Exploration (IGME).

Jelen, J., 2018. Mining heritage and mining tourism. *Czech Jorunal of Tourism*, 7 (1), 93–105. DOI: 10.1515/cjot-2018–0005.

Jolliffe, I., Conlin, M., 2015. Lessons in transforming mines into tourism attractions. Chapter 20 in book. *Mining Heritage and Tourism: A Global Synthesis*, M.V. Conlin and L. Jolliffe (editors), Routledge, Taylor & Francis Group, 1st edition, pp. 485–505. https://doi.org/10.4324/9780203865507|

Jones, A.L., Flynn, K., 2015. Flogging a dead horse or creating cultural capacity? The development and impact of mines as alternative tourist destinations: A case study of. Chapter 11 in book. *Mining Heritage and Tourism: A Global Synthesis*, M.V. Conlin and L. Jolliffe (editors), Routledge, Taylor & Francis Group, 1st edition, https://doi.org/10.4324/9780203865507

Jones, C., Munday, M., 2001. Blaenavon and united nations world heritage site status: Is conservation of industrial heritage a road to local economic development? *Regional Studies*, 35 (6), 585–590. https://doi.org/10.1080/00343400120065741

Jonsen-Verbeke, M., 1999. Industrial heritage: A nexus for sustainable tourism development. *Tourism Geographies*, 1 (1), 70–85. https://doi.org/10.1080/14616689908721295

Kaçar, A.D., 2016. Learning from the Ruhr: The case of the world heritage site Zollverein as a model of conserving industrial culture in Turkey. *Kent Araştırmaları Dergisi (Journal of Urban Studies)*, 7 (19), 474–496. Available via: <https://dergipark.org.tr/tr/pub/idealkent/issue/36795/419446>

Kakavogianne, O., Douni, K., Georgakopoulou, M., 2019. The cupellation of argentiferous lead in Mesogeia, East Attica, during prehistoric times. Proceedings of the International Conference "Ari and the Laurion from Prehistoric to Modern Times", Ruhr-Universität Bochum, Germany, 11, 3–4. (Quoted from Voudouris, 2021).

Kakavoyannis, E.Ch., 2005. *Μέταλλα εργάσιμα και συγκεχωρημένα – Η οργάνωση της εκμετάλλευσης του ορυκτού πλούτου της Λαυρεωτικής από την Αθηναϊκή Δημοκρατία*, TAPA. Quoted from (Kayafa, 2018).

Kakavoyannis, E.Ch., Koursoumis S.S., 2013. Εντοπισμός, καταγραφή και χαρτογράφηση των αρχαίων μεταλλευτικών φρεάτων της Λαυρεωτικής. *Archaeologiki Ephemeris*, 152, 77–102. Quoted from (Kayafa, 2018).

Kaliambakos, D., 2015. *The Contribution of the National Technical University of Athens in Lavrion of the 21st Century*, Athens University Publications NTUA. Quoted from (Chatzi Rodopoulou, 2020).

Kaliampakos, D., Benardos, A., 2009a. Developing the first underground hazardous waste repository in Greece. Proceedings of the 3rd International Conference, September 7–9, pp. 182–187.

Kaliampakos, D., Benardos, A., 2009b. On-site underground hazardous waste repositories: A successful model. 12th ACUUS International Conference, Shenzhen, China, November 18–19, pp. 14–18.

Kaliampakos, D., Damigos, D., Karachaliou, T., 2007. Using the "dry tomb" technique in the remediation of heavily contaminated land. *Proceedings of the International Symposium on Environmental Issues and Waste Management in Energy and Mineral Production – SWEMP 2007*, December 11–13. Quoted from (Benardos and Kaliampakos, 2013).

Kaliampakos, D., Damigos, D., Protonotarios, V., 2009. Management of a highly contaminated structure resulted from metallurgical activity: The rare case of 'Konofagos Building'. Proceedings of the 3rd AMIREG International Conference, September 7–9, pp. 188–193. Quoted from (Benardos and Kaliampakos, 2013).

Kapetanios, A., 2013. Ο χώρος και οι άνθρωποι στην αρχαία τοπογραφία της Λαυρεωτικής. *Αρχαιολογικές Συμβολές, τόμος Α: Αττική, ΚΣΤ' και Β' Εφορείες Προϊστορικών και Κλασικών Αρχαιοτήτων*, M. Donga-Toli and S. Economou (editors), Museum of Cycladic Art, pp. 183–198. Quoted from (Kayafa, 2018).

Karachaliou, T., Kaliampakos, D., 2005. Redeveloping derelict urban space, the case study of Lavrio, Greece. Proceedings of the International Conference on Managing Urban Land CABERNET 2005, April 13–15. Quoted from (Benardos and Kaliampakos, 2013).

Karadağ, A., İncedere, L., 2020. Kentsel belleğin sürdürülebilirliği açısından İzmir'deki endüstri miras alanlarının önemi: Alsancak Liman Ardı Bölgesi örneği (Importance of industrial heritage areas in İzmir in terms of sustainable urban memory: Case of hinterland of Alsancak Port. *Aegean Geographical Journal*, 29 (1), 57–71. Available via: <https://dergipark.org.tr/tr/pub/ecd/issue/55073/732982>

Kariptaş, F.S., Edirne Erdinç, J., Özkanç Dinçer, B., 2015. A study of industrial heritage in the context of cultural sustainability in urban areas. *2nd International Sustainable Building Symposium*, 28–30 May, pp. 512–516.

Katerinopoulos, A., 2010. The Lavrion mines. *Natural Heritage from East to West*, N. Evelpidou, T. Figueiredo, F. Mauro and A. Vassilopoulos (editors), Springer-Verlag, National and Kapodistrian University of Athens, Department of Geology and Geoenvironment, Section of Mineralogy and Petrology, Panepistimiopolis, Gr-15784, pp. 27–33. https://doi.org/10.1007/978-3-642-01577-9_3

Katerinopoulos, A., Zissimopoulou, E., 1994. *Minerals of the Lavrion Mines*, The Greek Association of Mineral and Fossil Collectors, pp. 1–304. Quoted from (Voudouris et al., 2021).

Kaya, M., Yılmaz, C., 2018. Endüstriyel miras turizmi için bir örnek: Ayancık – Zingal Orman İşletmesi (Sinop). (An example for industrial heritage tourism: Ayancık – Zingal Forest Enterprise (Sinop)). *Erciyes University Journal of Social Sciences Institute*, 44 (1), 121–162. Available via: <https://dergipark.org.tr/tr/download/article-file/511182>

Kayafa, M., 2018. Multiple readings of the mining landscape in Lavreotiki (SE Attica, Greece). *Bulletin of the Geological Society of Greece*, 53 (1), 99–124. https://doi. org/10.12681/bgsg.18639

Kazancı, N., Erdem, N.Ö., Erturaç, M.K., 2017. Cultural geology and geological heritage; new initiatives for earthsciences. *Geological Bulletin of Turkey*, 60 (1), 1–16. https://doi.org/10.25288/tjb.297797

Kılıç Yıldız, Ş., 2021. Industrial heritage making in Britain: The conservation, adaptation, and reinterpretation of the historic textile industry complexes. *Journal of Humanities and Tourism Research*, 11 (4), 702–721. https://doi.org/10.14230/johut1114

Kilinc, G., Gulersoy, N.Z., 2011. Evaluation of the potential for districts/counties to become provinces with respect to the level of urbanization in Turkey. *European Planning Studies*, 19 (8), 1557–1580. https://doi.org/10.1080/09654313.2011.586176

Kilinc, N., Turk, S.S., 2021. Plan changes in Istanbul (Turkey) as project-led practices in a plan-led planning system. *European Planning Studies*, 29 (8), 1393–1418. https://doi.org/10.1080/09654313.2020.1865276

Klemme, S., Berndt, J., Mavrogonatos, C., Flemetakis, S., Baziotis, I., Voudouris, P., Xydous, S., 2018. On the Color and Genesis of Prase (Green Quartz) and Amethyst from the Island of Serifos, Cyclades, Greece. *Minerals*, 8, 487. (Voudouris vd., 2021).

Knapp, A.B., Ashmore, W., 1999. Archaeological landscapes: Constructed, conceptualized, ideational. *Archaeologies of Landscape: Contemporary Perspectives*, W. Ashmore and A.B. Knapp (editors), Blackwell, pp. 1–30. Quoted from (Kayafa, 2018).

Knight, J., 2018. Transforming the physical geography of a city: An example of Johannesburg, South Africa. Chapter 8 in "Urban Geomorphology". *Landforms and Processes in Cities*, pp. 129–147. https://doi.org/10.1016/B978-0-12-811951-8.00008-4

Kömürlü, E., Kesimal, A., 2016. Rock bolting from past to present in 20 inventions. *MT Bilimsel, Journal of Underground Resources*, 5 (9), 69–85. Available via: <https://dergipark.org.tr/en/pub/mtb/issue/32055/354908>

Koramaz, T.K., Gülersoy, N.Z., 2011. Users' responses to 2D and 3D visualization techniques, representing the change in historic townscape. *Disegnarecon*, 7 (4), 30–41. https://doi.org/10.6092/issn.1828-5961/2299

Kossoff, D., Hudson-Edwards, K.A., Howard, A.J., Knight, D., 2016. Industrial mining heritage and the legacy of environmental pollution in the Derbyshire Derwent catchment: Quantifying contamination at a regional scale and developing integrated strategies for management of the wider historic environment. *Journal of Archaeological Science: Reports*, 6, 190–199. https://doi.org/10.1016/j.jasrep.2016.02.007

Koumantakis, I., Panagopoulos, A., Markantonis, K., 1999. *Geological and Hydrogeological Investigation in the Area of LTCP*. Report (in Greek).

Kretschmann, J., 2013. Stakeholder orientated sustainable land management: The Ruhr Area as a role model for urban areas. *International Journal of Mining Science and Technology*, 23 (5), 659–663. https://doi.org/10.1016/j.ijmst.2013.08.007

Królikowska, K., 2017. Narzędzia wdrażania zasad turystyki zrównoważonej na obszarach górskich. *Rozprawy naukowe Akademii Wychowania Fizycznego we Wrocławiu*, 56, 36–51 (in Polish). Quoted from (Gałaś et al., 2018c).

Labadi, S., 2018. Historical, theoretical and international considerations on culture, heritage and (sustainable) development. Chapter 3 in book: *World Heritage and Sustainable Development: New Directions in World Heritage Management*, P.B. Larsen and W. Logan (editors), Routledge, Taylor & Francis Group, 1st edition. https://doi.org/10.4324/9781315108049

Lageard, J.G.A, Drew, I.B., 2015. Evaporating Legacies: Industrial heritage and salt in cheshire, UK. *Industrial Archaeology Review*, 37 (1), 48–61. https://doi.org/10.1179 /0309072815Z.00000000042

Landford, C., 2009. Managing for sustainable tourism: A review of six cultural World Heritage Sites. *Journal of Sustainable Tourism*, 17 (1), 53–70. https://doi. org/10.1080/09669580802159719

Lane, B., Weston, R., Davies, N., Kastenholz, E., Lima, J., Majewsjki, J., 2013. Industrial heritage and agri/rural tourism in Europe. *European Parliament's Committee*, European Union. https://doi.org/10.2861/4530

Laraia, M., 2019. The fundamentals of industrial redevelopment. Chapter 2. *Beyond Decommissioning: The Reuse and Redevelopment of Nuclear Installations Woodhead Publishing Series in Energy*, ScienceDirect, pp. 15–58. https://doi.org/10.1016/ B978-0-08-102790-5.00002-6

Larsen, P.B., 2018. Human rights, wrongs and sustainable development in World Heritage. Chapter 8 in book. *World Heritage and Sustainable Development: New Directions in World Heritage Management*, P.B. Larsen and W. Logan (editors), Routledge, Taylor & Francis Group, 1st edition. https://doi.org/10.4324/9781315108049

Lemky, K., Jolliffe, L., 2015. Mining heritage and tourism in the former coal mining communities of Cape Breton Island, Canada. Chapter 12 in book. *Mining Heritage and Tourism: A Global Synthesis*, M.V. Conlin and L. Jolliffe (editors), Routledge, Taylor & Francis Group, 1st edition. https://doi.org/10.4324/9780203865507|

Lewis, I.D., 2020. Linking geoheritage sites: Geotourism and a prospective Geotrail in the Flinders Ranges World Heritage Nomination area, South Australia. *Australian Journal of Earth Sciences*, 67 (8), 1195–1210. https://doi.org/10.1080/08120099.2 020.1817147

Lichrou, M., O'Malley, L., 2006. Mining and tourism: Conflicts in the marketing of Milos Island as a tourism destination. *Tourism and Hospitality Planning & Development*, 3 (1), 35–46. https://doi.org/10.1080/14790530600640834

Liesch, M., 2014. Spatial boundaries and industrial landscapes at Keweenaw National Historical Park. *The Extractive Industries and Society*, 1 (2), 303–311. https://doi. org/10.1016/j.exis.2014.08.007

Lin, H., 2013. Conservation and community consciousness. Chapter 21 in book. *Industrial Heritage Re-Tooled: The TICCIH Guide to Industrial Heritage Conservation*, J. Douet (editor), Routledge, Taylor & Francis Group, 1st edition. https://doi. org/10.4324/9781315426532

Lin, Y.C., Liu, Y.C., 2018. Deconstructing the internal structure of perceived authenticity for heritage tourism. *Journal of Sustainable Tourism*, 26 (12), 2134–2152. https:// doi.org/10.1080/09669582.2018.1545022

Logan, W., 2018. Heritage, Sustainable Development and the achievement of peace and security in our world. Chapter 9 in book. *World Heritage and Sustainable Development: New Directions in World Heritage Management*, P.B. Larsen and W. Logan (editors), Routledge, Taylor & Francis Group, 1st edition. https://doi. org/10.4324/9781315108049

Logan, W., Larsen, P.B., 2018. Policy-making at the World Heritage-sustainable development interface. Chapter 1 in book. *World Heritage and Sustainable Development: New Directions in World Heritage Management*, P.B. Larsen and W. Logan (editors), Routledge, Taylor & Francis Group, 1st edition. https://doi. org/10.4324/9781315108049

MacLeod, N., 2016. Self-guided trails – A route to more responsible tourism? *Tourism Recreation Research*, 41 (2), 134–144. https://doi.org/10.1080/02508281.2016.114 7222

Madden, M., Shipley, R., 2012. An analysis of the literature at the nexus of heritage, tourism, and local economic development. *Journal of Heritage Tourism*, 7 (2), 103–112. https://doi.org/10.1080/1743873X.2011.632483

Manteca Martinez, J.I., 2013. The situation of the geo-mining heritage in the Region of Murcia and mining tourism as a sustainable alternative for its protection. *Regional and National Current Status, Seminario SEM (Sociedad Espanola de Mineralogia)*, 10, 62–74. Available via: <www.semineral.es/websem/PdfServlet?mod=archivos&s ubMod=publicaciones&archivo=SeminSEMv10p62-74.pdf>

Mapesa, M.W., 2018. Heritage: Conservation vs development – Challenging our and attitudes, February 7. Available via: <https://vdocuments.mx/heritage-conservation-vs-development-cro-15-papers-web-viewheritage-conservation.html?page=2>

Marinos G., Petrascheck, W.E., 1956. *Λαύριον*. Quoted from (Kayafa, 2018).

Markouli, A., 2019. 19th/20th Cent. Mining and metallurgy in Lavrion. Proceedings of the International Conference "Ari and the Laurion from Prehistoric to Modern Times", Ruhr-Universität Bochum, 11, p. 14. Quoted from (Voudouris et al., 2021).

Marot, N., Harfst, J., 2021. Post-mining landscapes and their endogenous development potential for small- and medium-sized towns: Examples from Central Europe. *The Extractive Industries and Society*, 8 (1), 168–175. https://doi.org/10.1016/j. exis.2020.07.002

Martin, P., 2013. Industrial archaeology. Chapter 5 in book. *Industrial Heritage Re-Tooled: The TICCIH Guide to Industrial Heritage Conservation*, J. Douet (editor), Routledge, Taylor & Francis Group, 1st edition. https://doi. org/10.4324/9781315426532

Mata-Perelló, J., Carrión, P., Molina, J., Villas-Boas, R., 2018. Geomining heritage as a tool to promote the social development of rural communities. Chapter 9 in book. *Geoheritage: Assessment, Protection, and Management*, pp. 167–177. https://doi. org/10.1016/B978-0-12-809531-7.00009-5

Mathias, R., 2018. Regional identity in the making? Industrial heritage and regional identity in the coal region of Northern Kyūshū in Japan. Chapter 7 in book. Industrial Heritage and Regional Identities, C. Wicke, S. Berger and J. Golombek (editors), *Routledge, Taylor & Francis Group*, 1st edition. https://doi.org/10.4324/9781315281179

Measham, T., Walton, A., Felton, S., 2021. Mining heritage and community identity in the social licence of proposed renewed mining. *The Extractive Industries and Society*, 8 (3), 100891. https://doi.org/10.1016/j.exis.2021.02.011

Metsaots, K., Printsmann, A., Sepp, K., 2015. Public opinions on oil shale mining heritage and its tourism potential. *Scandinavian Journal of Hospitality and Tourism*, 15 (4), 380–399. https://doi.org/10.1080/15022250.2015.1024817

Mieg, H.A., Oevermann, H., Noll, H.P., 2020. Conserve and innovate simultaneously? Good management of European UNESCO industrial world heritage sites in the context of urban and regional planning. *disP – The Planning Review*, 56 (3), 20–33. https://doi.org/10.1080/02513625.2020.1851903

Migoń, P., 2018. Geoheritage and world heritage sites. *Geoheritage, Assessment, Protection and Management*, E. Reynard and J. Brilha (editors), Elsevier, pp. 237–249. Quoted from (Periferakis et al., 2019).

Migone, J., 2013. Heritage at risk surveys. Chapter 16 in book. *Industrial Herit-age Re-Tooled: The TICCIH Guide to Industrial Heritage Conservation*, J. Douet (editor), Routledge, Taylor & Francis Group, 1st edition. https://doi. org/10.4324/9781315426532

Mills, C., McIntosh, I., 2021. 'I see the site of the old colliery every day': Scotland's landscape legacies of coal. *Landscapes*, 21 (1), 50–71. https://doi.org/10.1080/1466 2035.2020.1864095

Moraiti, E., 2015. Geotopes-Geoparks in Greece, contribution to sustainable develop-ment European Geopark Conference. Proceedings of the European Geopark Confer-ence, Rokua, Finland, 3–6 September, p. 144. Quoted from (Voudouris et al., 2021).

Moraiti, E., Staridas, S., 2015. Lavrion: One great territory with unique geoheritage as a prospective Geopark. Proceedings of the European Geopark Conference, Rokua, Finland, 3–6 September, p. 147. Quoted from (Voudouris et al., 2021).

Morin, B., 2013. Financial and fiscal incentives. Chapter 15 in book. *Industrial Heritage Re-Tooled: The TICCIH Guide to Industrial Heritage Conservation*, J. Douet (editor), Routledge, Taylor & Francis Group, 1st edition. https://doi. org/10.4324/9781315426532

Morin, D., Delpech, S., 2018. Mines and mining. *The Section of Mediterranean Archae-ology, Department of Archaeology*, Exploring Thorikos; R.F. Docter and M. Webster (editors), Ghent University, pp. 41–43. Quoted from (Voudouris et al., 2021).

Morin, D., Photiades, A., 2008. L'exploitation des gisements métallifères profonds dans l'Antiquité. Les mines du Laurion (Grèce). *Die Eckart Olshausen/Vera Sauer (Hg) Die Schätze der Erde – Natürliche Ressourcen in der antiken Welt*, E. Olshausen and V. Sauer (editor), Stuttgarter Kolloquium zur Historischen Geographie des Al-tertums 10, Geographica Historica 28; CD-Rom Illustrations; Franz Steiner Verlag, pp. 28–335. Quoted from (Voudouris et al., 2021).

Morin, D., Rosenthal, P., Photiades, A., Delpech, S., Jacquemot, D., 2020. Aegean mining technologies in antiquity – A traceological approach: The Laurion Mines (Greece). *Metallurgy in Numismatics 6 – Mines, Metals, and Money Ancient World Studies in Science, Archaeology and History*, K.A. Sheedy and G. Davis (editors), The Royal Numismatic Society Special Publication, pp. 23–41. Quoted from (Vou-douris et al., 2021).

Moroni, A., Gnezdilova, V.V., Ruban, D.A., 2015. Geological heritage in archaeologi-cal sites: case examples from Italy and Russia. *Proceedings of the Geologists' As-sociation*, 126 (2), 244–251. https://doi.org/10.1016/j.pgeola.2015.01.005

Negri, M., 2013. Industrial museums. Chapter 25 in book. *Industrial Heritage Re-Tooled: The TICCIH Guide to Industrial Heritage Conservation*, J. Douet (editor), Rout-ledge, Taylor & Francis Group, 1st edition. https://doi.org/10.4324/9781315426532

Niemiec, D., Duraj, M., Marschalko, M., Yilmaz, I., 2016. Conservation of se-lected churches in the most region and Karviná region and their significance for geotourism. *Procedia Engineering*, 161, 2276–2281. https://doi.org/10.1016/j. proeng.2016.08.828

NTUA, 1997. *Technological and Cultural Park of Lavrion: Studies and Works 1994–97*, University publications NTUA, Quoted from (Chatzi Rodopoulou, 2020).

Oakley, P., 2018. After mining: Contrived dereliction, dualistic time and the moment of rupture in the presentation of mining heritage. *The Extractive Industries and Society*, 5 (2), 274–280. https://doi.org/10.1016/j.exis.2018.03.005

Oevermann, H., Mieg, H.A., 2017a. Transformations of industrial heritage sites: Heritage and planning. Chapter 1. *Industrial Heritage Sites in Transformation: Clash of Discourses*, H.A. Mieg and H. Oevermann (editors), Routledge, Taylor & Francis Group, 1st edition. https://doi.org/10.4324/9781315797991

Oevermann, H., Mieg, H.A., 2017b. Studying transformations of industrial heritage sites: Synchronic discourse analysis of heritage conservation, urban development, and architectural production. Chapter 2 in book. *Industrial Heritage Sites in Transformation: Clash of Discourses*, H.A. Mieg and H. Oevermann (editors), Routledge, Taylor & Francis Group, 1st edition. https://doi.org/10.4324/9781315797991

Oglethorpe, M., McDonald, M., 2013. Recording and documentation. Chapter 7 in book. *Industrial Heritage Re-Tooled: The TICCIH Guide to Industrial Heritage Conservation*, J. Douet (editor), Routledge, Taylor & Francis Group, 1st edition. https://doi.org/10.4324/9781315426532

Okada, M., 2013. Industrial ruins. Chapter 20 in book. *Industrial Heritage Re-Tooled: The TICCIH Guide to Industrial Heritage Conservation*, J. Douet (editor), Routledge, Taylor & Francis Group, 1st edition. https://doi.org/10.4324/9781315426532

Önkol, G., 2012. Dönüşen kentlerde değişen kimlikler: Toplumsal bellek ve özgürlük yitimi. (Changing identities in transforming cities: Social memory and loss of freedom) *Mimarist Dergisi (Architects Journal)*, 43, 101–107.

Ostręga, A., Cała, M., Sinkovic, J., Kaleta, M., Mylona, R., Román-Ross, G., Bruno – Amphos, A., Amaro, S.L., Barbarosa, S., Ammerer, G., 2020. D4.2 *Elaborating Models of Revitalization in the Postmining Region*. Revitalising Post Mining Regions (ReviRis), EIT Raw Materials. Available via: <https://haldus.taltech.ee/sites/default/files/2021-12/D4.1_Development%20Model_0.pdf>

Ottens, B., Voudouris, P., 2018. *Griechenland: Mineralien-Fundorte-Lagerstätten*, Christian Weise Verlag, p. 480.

Oygür, A.V., 2021. Madencilik mirası ve maden turizmi (Mining heritage and mining tourism). *Mining Turkey Magazine*, 98, 98–104. Available via: <https://madencilikturkiye.com/wp-content/uploads/2018/09/Madencilik-Turkiye-Dergisi-Sayi-98-asd-jhsa73rozc4r.pdf>

Özen, H., Sert, A., 2006. Karadeniz'de unutulan endüstri mirasi (The forgotten industrial heritage in Black Sea Region). *Journal of the Faculty of Engineering and Architecture of Gazi University*, 21 (3), 499–508. Available via: <https://dergipark.org.tr/en/pub/gazimmfd/issue/6669/88837>

Ozus, E., Turk, S.S., Dokmeci, V., 2011. Urban restructuring of Istanbul. *European Planning Studies*, 19 (2), 331–356. https://doi.org/10.1080/09654313.2010.51582

Panagopoulos, I., Karayannis, A., Adam, K., Aravossis, K., 2007. Project and risk management for the reclamation of old mine sites. Proceedings of the International Conference of Environmental Management, Engineering, Planning and Economics, June 24–28, pp. 667–673.

Panagopoulos, K.I., 2008. Ποια μπορεί να είναι η συμβολή του Τεχνολογικού Πάρκου Λαυρίου στην πόλη του Λαυρίου και στην ευρύτερη περιοχή της Λαυρεωτικής. ΙnΠρακτικά Θ' Επιστημονικής Συνάντησης ΝΑ Αττικής, Λαύριο Αττικής 13–16 Απριλίου, 691–693. Quoted from (Kayafa, 2018).

Papadimitriou, V., 2012. Εκπαίδευση βασισμένη στον "Τόπο". Ηλεκτρονικό περιοδικό της ΠΕΕΚΠΕ. *Τεύχος*, 1 (46). Quoted from (Kayafa, 2018).

Papaspyridakou, P., 2013. Type-locality and aesthetic minerals from Lavrion: A new geopath for their quest. Master's thesis, National and Kapodistrian University of Athens, Greece. Quoted from (Voudouris et al., 2021).

Papaspyridakou, P., Katerinopoulos, A., Voudouris, P., Magganas, A., Megremi, I., 2017. A geopath in a quest for aesthetic minerals at Lavrion mines. Sofia Initiative "Mineral diversity preservation". Proceedings of the IX International Symposium Mineral Diversity Research and Reservation, Sofia, Bulgaria, 16–18 October, p. 47. Quoted from (Voudouris et al., 2021).

Peng, K.H., Tzeng, G.H., 2019. Exploring heritage tourism performance improvement for making sustainable development strategies using the hybrid-modified MADM model. *Current Issues in Tourism*, 22 (8), 921–947. https://doi.org/10.1080/136835 00.2017.1306030

Perelli, C., Pinna, P., Sistu, G., 2015. Mining heritage, local development and territory identity: The case of Sardinia. Chapter 17 in book. *Mining Heritage and Tourism: A Global Synthesis*, M.V. Conlin and L. Jolliffe (editors), Routledge, Taylor & Francis Group, 1st edition, https://doi.org/10.4324/9780203865507|

Perfetto, M.C., Vargas-Sánchez, A., 2017. Towards a smart tourism business ecosystem based on industrial heritage: Research perspectives from the mining region of Rio Tinto, Spain. *Journal of Heritage Tourism*, 13 (6), 528–549. https://doi.org/10.1080 /1743873X.2018.1445258

Periferakis, A., Paresoglou, I., Paresoglou, N., 2019. The significance of the Lavrion mines in Greek and European Geoheritage. *European Geologist*, 48, 24–27. Available via: <www.researchgate.net/publication/337389596_The_significance_of_the_ Lavrion_mines_in_Greek_and_European_Geohe ritage>

Periferakis, A., Paresoglou, N., 2019. Lavrion from Ancient Greece to the present day: A study of how an ore deposit shaped history. 15th International Congress of the Geological Society of Greece, 22–24 May, Athens, Greece, pp. 704–705. Available via: <www.researchgate. net/publication/333782810_Lavrion_from_ Ancient_Greece_to_the_Present_Day_A_Study_of_how_an_Ore_Deposit_ Shaped_History>

Pfaffenberger, B., 1998. Mining communities, chaînes opératoires and sociotechnical systems. *Social Approaches to an Industrial Past – The Archaeology and Anthropology of Mining*, A.B. Knapp, V.C. Pigott and E.W. Herbert (editors), Routledge, pp. 291–300. Quoted from (Kayafa, 2018).

Photos-Jones, E., Ellis Jones, J., 1994. The building and industrial remains at Agrileza, Laurion (fourth century BC) and their contribution to the workings at the site. *Annual of the British School at Athens*, 89, 307–358. Quoted from (Kayafa, 2018).

Pinto, B.B., 2014. Fugger – Agricola – Hoover. *Turkish Miners Association (TMD), Sector News Bulletin*, 50, 58–60. Available via: <www.tmder.org.tr/modules/faq/da-tafiles/FILE_FE76CC-A9D27D-8321A6-45A0C6-014D4F-97BD10.pdf>

Pogkas, K., 1996. How did we get to the Lavrion Technological and Cultural Park *Sigchrona Themata*, 58–59, 33–35. Quoted from (Chatzi Rodopoulou, 2020).

Polyxen, M., Spyridon, P., n.d. Representations and perceptions of industrial heritage in the higher technological education institute in Greece: An empirical study. *E-Journal of Science & Technology*, 33–48.

Pop, D., Horák, J., Hurlbut, J.M., 2004. Mineral museums as alliance partners in teaching mineral sciences. *Journal of Geoscience Education*, 52 (1), 87–96. https://doi. org/10.5408/1089-9995-52.1.87

Power, R., 2008. "After the black gold": A view of mining heritage from coalfield areas in Britain. *Folklore*, 119 (2), 160–181. https://doi.org/10.1080/00155870802056977

Preite, M., 2013. Urban regeneration and planning. Chapter 13 in book. *Industrial Heritage Re-Tooled: The TICCIH Guide to Industrial Heritage Conservation*,

J. Douet (editor), Routledge, Taylor & Francis Group, 1st edition. https://doi.
org/10.4324/9781315426532

Price, W.R., Rhodes II, M.A., 2022. Coal dust in the wind: Interpreting the industrial
past of South Wales. *Tourism Geographies*, 24 (4–5), 837–858. https://doi.org/10.10
80/14616688.2020.1801825

ProGeo, 2011. Conserving our shared geoheritage – A protocol on geoconservation
principles, sustainable site use, management, fieldwork, fossil and mineral collecting. Available via: <www.progeo.se /progeo-protocol-definitions-20110915.pdf>

Prosser, C.D., 2019. Communities, quarries and geoheritage – Making the connections.
Geoheritage, 11 (4), 1–13. https://doi.org/10.1007/s12371-019-00355-4

Protonotarios, V., Petsas, N., Moutsatsou, A., 2002. Levels and composition of atmos-
pheric particulates (PM10) in a mining–industrial site in the City of Lavrion, Greece.
Journal of The Air & Waste Management Association, 52, 1263–1273.

Reeves, K., McConville, C., 2011. Cultural landscape and goldfield heritage: Towards a
land management framework for the historic South-West Pacific gold mining landscapes.
Landscape Research, 36 (2), 191–207. https://doi.org/10.1080/01426397.2010.547573

Rehren, Th., Vanhove, D., Mussche, H., 2002. Ores from the ore washeries in Lavreotiki. In Metalla. *Bochum* 9 (1), 27–46. Quoted from (Kayafa, 2018).

Rentzhog, S., 2007. *Open Air Museums. The History and Future of a Visionary Idea*,
Kristianstad.

Rezafar, A., Turk, S.S., 2016. The history of aesthetic control and management in
the planning system, the case of Turkey. International Planning History Society Proceedings, 17th IPHS Conference, History-Urbanism-Resilience, TU Delft
17–21 July 2016,V.06, pp. 165–175, TU Delft Open. http://dx.doi.org/10.7480/
iphs.2016.6.1327

Rezafar, A., Turk, S.S., 2018. Urban design factors involved in the aesthetic assessment of newly built environments and their incorporation into legislation:
The case of Istanbul. *Urbani Izziv*, 29 (2), 83–95. http://dx.doi.org/10.5379/
urbani-izziv-en-2018-29-02-002

Rezafar, A., Turk, S.S., 2019. Kentsel estetiği etkileyen faktörlerin belirlenmesi ve
farklı aktörler tarafından değerlendirilmesi. Türkiye Kentsel Morfoloji Araştırma
Ağı. 2. Kentsel Morfoloji Sempozyumu: "DeğişKent" Değişen Kent, Mekân ve
Biçim, s.161–172, Istanbul Teknik üniversitesi, Taşkışla.

Rhatigan, J., 2020. Mining meaning: telling spatial histories of the Britannia Mine. *Journal of Historical Geography*, 67, 36–47. https://doi.org/10.1016/j.jhg.2019.10.010

Rieck, B., Lengauer, C.L., Giester, G., 2019. Select Voudourisite, Cd(SO4)·H2O, and
lazaridisite, Cd3(SO4)3·8H2O, two new minerals from the Lavrion Mining District, Greece. *Mineralogical Magazine*, 83 (4), 551–559. https://doi.org/10.1180/
mgm.2018.157 Quoted from (Periferakis et al., 2019).

Rieck, B., Rieck, P., 1999. Lavrion: The complete mineral list. *Lapis*, 24, 61–63. (In
German). Quoted from (Voudouris et al., 2021).

Roche, C., Sinclair, L., Spencer, R., Luke, H., Brueckner, M., Knowles, S., Paull, M.,
2021. A mining legacies lens: from externalities to wellbeing in extractive industries. *The Extractive Industries and Society*, 8 (3), 100961. https://doi.org/10.1016/j.
exis.2021.100961

Rodrigues, J., de Carvalho, C.N., Ramos, M., Ramos, R., Vinagre, A., Vinagre, H.,
2021. Geoproducts – Innovative development strategies in UNESCO Geoparks:
Concept, implementation methodology, and case studies from Naturtejo Global

Geopark, Portugal. *International Journal of Geoheritage and Parks*, 9 (1), 108–128. https://doi.org/10.1016/j.ijgeop.2020.12.003

Rodwell, D., 2017. Liverpool: Heritage and development – Bridging the gap? Chapter 3 in book. *Industrial Heritage Sites in Transformation: Clash of Discourses*, H.A. Mieg and H. Oevermann (editors), Routledge, Taylor & Francis Group, 1st edition. https://doi.org/10.4324/9781315797991

Ross, J., Voudouris, P., Melfos, V., Vaxevanopoulos, M., Soukis, K., Merigot, K., 2021. *What Did the Ancient Greeks Mine at Lavrion and When Did They Mine It?* Der Anschnitt.

Rybar, P., 2010. Assessment of attractiveness (value) of geotouristic objects. *Acta Geoturistica*, 1 (2), 13–21. Available via: <https://geotur.fberg.tuke.sk/pdf/2010/n02/02_Rybar_v1_n2.pdf>

Sadioğlu, U., Yürük, Ö., 2020. Eskişehir'in Endüstriyel Mirası Fabrikalar Bölgesinin Kent Kimliği Üzerindeki Etkileri (Effects of factories region, industrial heritage of Eskişehir on urban identity). *Kent Araştırmaları Dergisi (Journal of Urban Studies), (ISSN: 1307–9905), Urbanization and Economy Special Issue*, 11 (3), 1049–1072. https://doi.org/10.31198/idealkent.673775

Saliora-Oikonomakou, M., 2004. *Ο Αρχαίος Δήμος του Σουνίου, ιστορική και τοπογραφική επισκόπηση*, Toumbis Editions. Quoted from (Kayafa, 2018).

Sallam, E.S., Ponedelnik, A.A., Tiess, G., Yashalova, N.N., Ruban, D.A., 2018. The geological heritage of the Kurkur – Dungul area in southern Egypt. *Journal of African Earth Sciences*, 137, 103–115. https://doi.org/10.1016/j.jafrearsci.2017.10.012

Sánchez-Cortez, J.L., 2019. Conservation of geoheritage in Ecuador: Situation and perspectives. *International Journal of Geoheritage and Parks*, 7 (2), 91–101. https://doi.org/10.1016/j.ijgeop.2019.06.002

Saner, M., 2012. Endüstri mirası: Kavramlar, kurumlar ve Türkiye'deki yaklaşımlar (Industrial heritage: Concepts, institutions and approaches in Turkey). TMMOB Chamber of City Plannrs, *Planlama (Planning)*, 52, 53–66. Available via: <www.researchgate.net/publication/255686097_Endustri_Mirasi_Kavramlar_Kurumlar_ve_Turkiye%27deki_Yaklasimlar_Mehmet_Saner>

Sanjoy, R.M., Sadhukhan, K., Chakrabarty, P., 2021. Alternative use of abandoned mines for geotourism: A case study using geoinformatics. Chapter 9. *Modern Cartography Series*, vol. 10, 191–204. https://doi.org/10.1016/B978-0-12-823895-0.00004-X

Santucci, V.L., 2005. Historical perspectives on biodiversity and geodiversity. *"Geodiversity and Geoconservation", The George Wright Forum*, 22 (3), 29–34. Available via: <www.georgewright.org/223santucci.pdf>

Schaal, D., 2017. Museums and industrial heritage: History, functions, perspectives. Chapter 10 in book. *Industrial Heritage Sites in Transformation: Clash of Discourses*, H.A. Mieg and H. Oevermann (editors), Routledge, Taylor & Francis Group, 1st edition. https://doi.org/10.4324/9781315797991

Schieder, W., 2005. Real slave prices and the relative cost of slave labor in the Greco-Roman World. *Ancient Society*, 35, 1–17. Available via: <www.jstor.org/stable/pdf/44079857.pdf>

Serrano, E., 2007. Geodiversity. A theoretical and applied concept. *Geographica Helvetica*, 62 (3), 140–147. https://doi.org/10.5194/gh-62-140-2007

Shih, N.J., Lin, C.Y., 2019. The evolving urban fabric and contour of old mountain streets in Taiwan. *Tourism Geographies*, 21 (1), 24–53. https://doi.org/10.1080/14616688.2017.1388437

Singh, B.V.R., Sen, A., Verma, L.M., Mishra, R., Kumar, V., 2021. Assessment of potential and limitation of Jhamarkotra area: A perspective of geoheritage, geo park and geotourism. *International Journal of Geoheritage and Parks*, 9 (2), 157–171. https://doi.org/10.1016/j.ijgeop.2021.04.001

Singh, R.S., Ghost, P., 2021. Geotourism potential of coal mines: An appraisal of Sonepur-Bazari open cast project, India. *International Journal of Geoheritage and Parks*, 9 (2), 172–181. https://doi.org/10.1016/j.ijgeop.2021.02.007

Sinnett, D.E., Sardo, A.M., 2020. Former metal mining landscapes in England and Wales: Five perspectives from local residents. *Landscape and Urban Planning*, 193, 103685. https://doi.org/10.1016/j.landurbplan.2019.103685

Sirel, A., Çerkezoğlu, M., 2016. Sustainability route of reusing of the industrial buildings in the field of cultural heritage: Discussion of Golden Horn Region in Istanbul. *A+Arch Design International Journal of Architecture and Design*, 2 (3), 1–16. Available via: <https://dergipark. org.tr/tr/pub/aarch/issue/42329/509299>

Smith, N., 2001. *Heritage of Industry: Discovering New Zealand's Industrial History*, Raupo Publishing. Quoted from (Karadağ and İncedere, 2020).

Smith, W., 2015. A smaller history of ancient Greece: From the earliest times to the Roman conquest. *CreateSpace Independent Publishing Platform*. Available via: <https://web.archive.org/web/20080510152211/www.ellopos.net/elpenor/greek-texts/ancient-greece/history-of-ancient-greece.asp>

Spitaels, P., 1984. The Early Helladic Period in Mine no.3. *Thorikos VIII 1972/1976*, H.F. Mussche et al. (editors), pp. 151–174. Quoted from (Kayafa, 2018).

Stos-Gale, Z.A., 2014. Silver vessels in the Mycenaean Shaft Graves and their origin in the context of the metal supply in the Bronze Age Aegean. *Metalle der Macht, Frühes Gold und Silber, 17. bis 19. Oktober 2013 Internationale Tagung in Halle (Saale)*, H. Meller, R. Risch and E. Pernicka (editors), Band 11/I, pp. 183–208. Quoted from (Kayafa, 2018).

Stuart, I., 2013. Identifying industrial landscapes. Chapter 6 in book. *Industrial Heritage Re-Tooled: The TICCIH Guide to Industrial Heritage Conservation*, J. Douet (editor), Routledge, Taylor & Francis Group, 1st edition. https://doi.org/10.4324/9781315426532

Stuttard, D., 2013. *Parthenon, Power and Politics on the Acropolis*, The British Museum Press.

Sutherland, F., 2015. Community-driven mining heritage in the Cuyuna Iron Mining District: Past, present, and future projects. *The Extractive Industries and Society*, 2 (3), 519–530. https://doi.org/10.1016/j.exis.2015.04.003

Swope, K.K., Gregory, C.J., 2020. Hard-rock mining for precious metals in the U.S. Southwest: Considerations for developing historic contexts and archaeological research designs. *Journal of Southwestern Anthropology and History*, 86, 176–185. https://doi.org/10.1080/00231940.2020.1749779

Syafrini, D., Nurdin, M.F., Sugandi, Y.S., Miko, A., 2020. The impact of multiethnic cultural tourism in an Indonesian former mining city. *Tourism Recreation Research*, 45 (4), 511–525. https://doi.org/10.1080/02508281.2020.1757208

Syafrini, D., Nurdin, M.F., Sugandi, Y.S., Miko, A., 2022. Transformation of a coal mining city into a cultured mining heritage tourism city in Sawahlunto, Indonesia: A response to the threat of becoming a ghost town. *Tourism Planning & Development*, 19 (4), 296–315. https://doi.org/10.1080/21568316.2020.1866653

Tang, Y., Liang, Y., 2022. Staged authenticity and nostalgia of mining tourists in the Jiayang mining Geo-park of China. *Journal of Tourism and Cultural Change*, https://doi.org/10.1080/14766825.2022.2090259

Tapete, D., Cigna, F., 2017. InSAR data for geohazard assessment in UNESCO World Heritage sites: State-of-the-art and perspectives in the Copernicus era. *International Journal of Applied Earth Observation and Geoinformation*, 63, 24–32. https://doi.org/10.1016/j.jag.2017.07.007

Tempel, N., 2013. Post-industrial landscapes. Chapter 19 in book. *Industrial Heritage Re-Tooled: The TICCIH Guide to Industrial Heritage Conservation*, J. Douet (editor), Routledge, Taylor & Francis Group, 1st edition. https://doi.org/10.4324/9781315426532

Terkenli, Th.S., 1996. *Το πολιτισμικό τοπίο: γεωγραφικές προσεγγίσεις*, Papazisi Editions. Quoted from (Kayafa, 2018).

Theodossiou-Drandaki, E., 1996. *Criteria of Characterizing Geotopes. 1st Symposium on the Conservation of Geological-Geomorphological Heritage*, pp. 70–72. (In Greek). Quoted from (Voudouris et al., 2021).

Theodossiou-Drandaki, E., 2001. Geological framework for the selection of geotopes according the criteria of the International Association of Geosciences (IUES) and the European Society of conservation of geological and geomorphological Heritage (ProGEO). *Bulletin of the Geological Society of America*, 34, 795–803. Quoted from (Voudouris et al., 2021).

Thompson, M., 2014. Mining heritage and tourism: A global synthesis. *Journal of Tourism History*, 6 (1), 99–101. https://doi.org/10.1080/1755182X.2014.964485

TICCIH, 2003. *The Nizhy Tagil Charter for the Industrial Heritage*, The International Committee for the Conservation of Industrial Heritage (TICCIH). Available via: <www.icomos.org/18thapril/ 2006/nizhny-tagil-charter-e.pdf>

TICCIH, 2011. *Dublin Principles*. The International Committee for the Conservation of Industrial Heritage (TICCIH). Available via: <http://ticcih.org/about/about-ticcih/dublin-principles/>

Timcak, G.M., Rybár, P., Jablonská, J., 2015. Geotourism site development in Slovakia. Chapter 13 in book. *Mining Heritage and Tourism: A Global Synthesis*, M.V. Conlin and L. Jolliffe (editors), Routledge, Taylor & Francis Group, 1st edition. https://doi.org/10.4324/9780203865507|

Timothy, D.J., Boyd, S., 2002. *Heritage Tourism*, Printice Hall.

TMMOB, 2006. File 03. *TMMOB Chamber of Architects Ankara Branch*. Available via: <www.mimarlarodasiankara.org/dosya/dosya3.pdf> Quoted from (Kaya and Yılmaz, 2018).

Tombal Kara, T.D., 2020. Madenciliğin ve cevher hazırlamanın kısa tarihi (A brief history of mining and mineral processing). Chapter 5 in book *"Academic Studies in the Field of Science and Mathematics" (ISBN: 978-625-7912-04-4)*, Gece Bookstore Publishing, Istanbul, pp. 81–102. Available via: <https://www.gecekitapligi.com/Webkontrol/uploads/Fck/FenBilimleriveMatematik.pdf>

Touliatos, P., Efesiou, I., 2010. 1992–2010 – From the initial design to the restoration and reuse of the historic buildings in the Lavrion Technological and Cultural Park. *Architectural Footprints of Industrial Archaeology in Lavrion from Documentation to Reuse Athens: University Publications NTUA*, Y. Tsilis (editor). Quoted from (Chatzi Rodopoulou, 2020).

Trinder, B., 2013. Industrial archaeology: A discipline? Chapter 3 in book. *Industrial Heritage Re-Tooled: The TICCIH Guide to Industrial Heritage Conservation*, J. Douet (editor), Routledge, Taylor & Francis Group, 1st edition. https://doi. org/10.4324/9781315426532

Tsaimou, C., Frangiskos, A., 2008. Ανάδειξη και αξιοποίηση του Αρχαίου Μεταλλευτικού Λαυρίου. Πρακτικά Θ' Επιστημονικής Συνάντησης ΝΑ Αττικής, Λαύρειο 13–16 Απριλίου, 647–659. Quoted from (Kayafa, 2018)

Tuğcu, A., 2012. Yesemek stone quarry and sculptural workshop. MSc. Thesis, Middle East Technical University.

Turk, S.S., 2002. The realization of the detailed local plans in urban areas in Turkey: A model. ERSA Conference Papers ersa02p408, European Regional Science Association. Available via: <https://ideas.repec.org/p/wiw/wiwrsa/ersa02p408.html>

Turk, S.S., Altes, W.K.K., 2010a. How suitable is LR for renewal of inner city areas? An analysis for Turkey. *Cities*, 27 (5), 326–336. https://doi.org/10.1016/j. cities.2010.03.010

Turk, S.S., Altes, W.K.K., 2010b. Potential application of land readjustment method (Lr) in urban renewal: An analysis for Turkey. *Journal of Urban Planning and Development*, 137 (1). https://doi.org/10.1061/(ASCE)UP.1943-5444.0000035

Türk, Ş.Ş., Ocakçı, M., Terzi, F., 2021. Kentsel dönüşüm sürecine planlama ilke ve kriterlerinin eklemlenmesi (Articulation of planning principles and criteria to the urban transformation process). *City and Planning: Hüseyin Kaptan'a Armağan*, YEM Publications, 1st edition, pp. 373–398.

Tzeferis, P., Bitzios, D., 2020. Lavrion Mines: On the threshold of accession as a UNESCO world heritage site? Available via: <www.sme.gr/metalleftiko-lavrio-stokatofli-entaksis-os-mnimeio-pagkosmias-klironomias-tis-unesco>

UNEP-WCMC, 2013. Identifying potential overlap between extractive industries (mining, oil and gas) and natural world heritage sites. Final Report. Available via: <www. unep-wcmc.org/resources-and-data/identifying-potential-overlap-between-extractive-industries-mining – oil-and-gas-and-natural-world-heritage-sites>

UNESCO, 2014. *Global Geoparks Network – Guidelines and Criteria for National Geoparks Seeking UNESCO's Assistance to Join the Global Geoparks Network*, UNESCO.

Urban Environment Laboratory, 2009. Architectural and museological design of Lavrion Metallurgy and Metallurgy Museum. *NTUA School of Architecture Urban Environment Laboratory*. Accessed date: 05/10/2017. Available via: <www. arch.ntua.gr/envlab/resources#resource-5310>. Quoted from (Chatzi Rodopoulou, 2020).

URL-1. Accessed date: 05/02/2017. Available via:

URL-2. A short history of Greece. Accessed date: 18/02/2022. Available via: <www. spainexchange.com/guide/GR-history.htm>

URL-3. LTCP & NTUA. Accessed date: 01/02/2022. Available via: < www.ltcp.ntua. gr/>

URL-4. List of natural history museums. Accessed date: 15/01/2022. Available via: <http://en.wikipedia.org/wiki/List_of_natural_history_museums>

URL-5. Laurium. Accessed date: 21/01/2022. Available via: <http://en.wikipedia. org/ wiki/Laurium>

URL-6. Accessed date: 18/04/2017. Available via: <www.lavrio-conferenceculturalpark.gr/52/article/english/52/43/index.htm>

URL-7. World Heritage Convention, UNESCO. Accessed date: 27/01/2022. Available via: <http://whc.unesco.org/en/tentativelist/>

URL-8. CruiseAway. Accessed date: 14/06/2020. Available via: <www.cruiseaway. com.au/cruise-port/lavrion-greece>

URL-9. Accessed date: 14/06/2018. Available via: <www.knowing project.eu/index>

URL-10. Accessed date: 21/09/2017. Available via: <www.patt.gov.gr/main/index. php?option=com_content&;view=article& knowing&catid>

URL-11. Official application letter to join the European and UNESCO Global Geoparks Networks. *Municipal Public Benefit Agency of Lavreotiki Kountoriotiki I*, 19500 Lavrio (pdf).

URL-12. Application Dossier for nomination as a European and Unesco Global Geopark. Accessed date: 18/07/2018. Available via: <http://old.igme.gr/Lavreotiki_Geopark_Application_Dossier.pdf>

URL-13. About LTCP. Accessed date: 11/01/2022. Available via: <https://en.ltcp.ntua. gr/about/>

URL-14. Greece – Lavrion Technological and Cultural Park. *Sabina*. Accessed date: 08/01/2022. Available via: <https://sabina-project.eu/lavrion-technological-and-cultural-park-greece/>

URL-15. Lavrion Technological and Cultural Park. *Yougoculture*. Accessed date: 10/01/2022. Available via: <www.yougoculture.com/virtual-tour/lavrin-sounion/experience/lavrion-technological-and-cultural-park>

URL-16. The Economy of Ancient Greece. Accessed date: 16/01/2022. Available via: <https://web.archive.org/web/20141022045544/>

URL-17. Ancient Greece. Accessed date: 17/01/2022. Available via: <https:// web.archive.org/web/20120113022308/www.olympia-greece.org/wiki/index. php?title=Ancient_Greece>

URL-18. Indo-European Chronology (IV period). Accessed date: 22/01/2022. Available via: <http://indoeuro.bizland.com/project/chron/chron3.html>

URL-19. The Ancient Greeks. Arrowhead Web Design & Consulting, Accessed date: 21/01/2022. Available via: <https://arwhead.com/Greeks/>

URL-20. Environmental Remediation Projects. Accessed date: 21/01/2022. Available via: <https://en.ltcp.ntua.gr/renovation/>

URL-21. LTCP n.d. Administration. Accessed date: 06/09/2017. Available: <www.ltp. ntua.gr/lavrion_park/administration_en> Quoted from (Chatzi Rodopoulou, 2020).

URL-22. History. Accessed date: 22/01/2022. Available via: <https://en.ltcp.ntua. gr/history>

URL-23. Culture. Available via: <https://en.ltcp.ntua.gr/ekdiloseis/>

URL-24. Technology & Innovation. Accessed date: 22/01/2022. Available via: <https:// en.ltcp.ntua.gr/technology-innovation/>

URL-25. Research-Education. Accessed date: 23/01/2022. Available via: <https:// en.ltcp.ntua.gr/research/>

URL-26. Research Projects. Accessed date: 23/01/2022. Available via: <https://en.ltcp. ntua.gr/projects/>

URL-27. Hosted Institutions. Accessed date: 23/01/2022. Available via: <https:// en.ltcp.ntua.gr/hosted_companies/>

URL-28. Event Venues. Accessed date: 24/01/2022. Available via: <https://en.ltcp.ntua. gr/venue/>

URL-29. Educational Activities. Accessed date: 24/01/2022. Available via: <https:// en.ltcp.ntua.gr/education/>

URL-30. Cinematic Park. Accessed date: 25/01/2022. Available via: <https://en.ltcp. ntua.gr/cinema/>

116   *References*

URL-31. Good practice: Lavrion Technological and Cultural Park (LTCP). *European Regional Development Fund.*

URL-32. The Colonization of Greece and Factors to Solons Reform in 594 B.C. *BookRags.* Accessed date: 15/01/2022. Available via: <www.bookrags.com/essay-2003/10/1/15849/2480/#gsc.tab=0>

URL-33. Protected Areas, Category III: Natural Monument or Feature. IUCN. Accessed date: 25/02/2022. Available via: <www.iucn.org/theme/protected-areas/about/protected-areas-categories/category-iii-natural-monument-or-feature>

URL-34. Tentative List, UNESCO. Accessed date: 11/09/2021. Available via: <https://whc.unesco.org/en/tentativelists/state=tr>

URL-35. World Heritage List. Accessed date: 17/02/2022. Available via: <https://whc.unesco.org/en/list/&order=country#alphaP>

URL-36. Zonguldak Mining Museum. Ministry of Culture and Tourism, Zonguldak Provincial Culture and Tourism Directorate. Accessed date: 28/02/2022. Available via: <https://zonguldak.ktb.gov.tr/TR-186810/zonguldak-maden-muzesi.html>

URL-37. Greek Mineral Wealth. Available via: <www.oryktosploutos.net /2019/08/a-trip-to-geological-and-ancient-mining/> Quoted from (Barbosa and Amaro, 2020).

URL-38. Adventure Mine: Experience the Salt Mines in Berchtesgaden. Available via: <www.berchtesgaden.de/en/salt-history/adventure-mine>

Uysal, M., 2018. Endüstriyel miras ve turizm ilişkisinin türkiye turizm planlaması kapsamında değerlendirilmesi (Evaluation of industrial heritage and tourism relationship on Turkey tourism planning). 1st International Conference on Tourism and Architecture, October 24–27, pp. 324–338. Available via: <www.researchgate.net/publication/335715539_Endustriyel_Miras_ve_Turizm_Iliskisinin_Turkiye_Turizm_Planlamasi_Kapsaminda_Degerlendirilmesi_-_Evaluation_of_Industrial_Heritage_and_Tourism_Relationship_on_Turkey_Tourism_Planning>

Vall, N., 2018. Coal is our strife: representing mining heritage in North East England. *Contemporary British History*, 32 (1), 101–120. https://doi.org/10.1080/13619462.2017.1408541

Vallianatou, E., 2012. Εθνικός Δρυμός Σουνίου και Λαύριο. Εισήγηση στο σεμινάριο του ΚΠΕ Λαυρίου: Η ιστορία και το περιβάλλον της πόλης μας οδηγεί στο μέλλον. Quoted from (Kayafa, 2018).

van Knippenberg, K., Boonstra, B., Boelens, L., 2022. Communities, heritage and planning: Towards a co-evolutionary heritage approach. *Planning Theory & Practice*, 23 (1), 26–42. https://doi.org/10.1080/14649357.2021.1998584

Vargas-Sánchez, A., 2015. Industrial heritage and tourism: A review of the literature. *The Palgrave Handbook of Contemporary Heritage Research*, E. Waterton and S. Watson (editors), Palgrave Macmillan, pp. 219–233.

Velitzelos, E., Mountrakis, D., Zouros, N., Soulakellis, N., 2003. *Atlas of the Geological Monuments of the Aegean; Ministry of the Aegean*, p. 352. (In Greek). Quoted from (Voudouris et al., 2021).

Voudouris, P., Katerinopoulos, A., Magganas, A., 2017. Mineralogical geotopes in Greece: Preservation and promotion of museum specimens of minerals and gemstones. Sofia Initiative "Mineral Diversity Preservation". *Proceedings of the IX International Symposium Mineral Diversity Research and Reservation*, Sofia, Bulgaria, October 16–18, pp. 149–159.

Voudouris, P., Mavrogonatos, C., Graham, I., Giuliani, G., Tarantola, A., Melfos, V., Karampelas, S., Katerinopoulos, A., Magganas, A., 2019a. Gemstones of Greece:

Geology and crystallizing environments. *Minerals*, 9, 461. https://doi.org/10.3390/min9080461

Voudouris, P., Mavrogonatos, C., Rieck, B., Kolitsch, U., Spry, P.G., Scheffer, C., Tarantola, A., Vanderhaeghe, O., Galanos, E., Melfos, V., Zaimis, S., Soukis, K., Photiades, A., 2018. The gersdorffite-bismuthinite-native gold association and the skarn-porphyry mineralization in the Kamariza Mining District, Lavrion, Greece. *Minerals*, 8 (11), 1–16. https://doi.org/10.3390/min8110531

Voudouris, P., Melfos, V., Mavrogonatos, C., Photiades, A., Moraiti, E., Rieck, B., Kolitsch, U., Tarantola, A., Scheffer, C., Morin, D., et al., 2021. The Lavrion Mines: A unique site of geological and mineralogical heritage. *Minerals*, 11, 76. https://doi.org/10.3390/min11010076

Voudouris, P., Melfos, V., Spry, P.G., Bonsall, T.A., Tarkian, M., Solomos, C., 2008. Carbonate-replacement Pb-Zn-Ag±Au in the Kamariza area, Lavrion, Greece: Mineralogical and thermochemical conditions of formation. *Mineralogy and Petrology*, 94, 85–106. https://doi.org/ 10.1007/s00710/-008–0007–4

Voudouris, P., Photiades, A., Tarantola, A., Scheffer, C., Vanderhaeghe, O., Morin, D., Vlachopoulos, N., 2019b. The Lavrion and Serifos mining centers: Two worldwide unique mineralogical and geological monuments and perspectives for their protection. Proceedings of the Greek ICOMOS Conference: "The Value Framework for the Protection and Management of Sites and Monuments Extracted during the Antiquity: Current Uses and Future Synergies", Athens-Lavrion, Greece, 29–30 November.

Vuin, A., Carson, D.A., Carson, D.B., Garrett, J., 2016. The role of heritage tourism in attracting "active" in-migrants to "low amenity" rural areas. *Rural Society*, 25 (2), 134–153. https://doi.org/10.1080/10371656.2016.1194324

Waybackmachine, 2009. Document about the end of the era of ancient Greece Quoted from (Wikipedia, 2022).

Wergeland, E.S., 2017. From high voltage to high density: The urban dynamism of cable street, Oslo. Chapter 4. *Industrial Heritage Sites in Transformation: Clash of Discourses*, H.A. Mieg and H. Oevermann (editors), Routledge, Taylor & Francis Group, 1st edition. https://doi.org/10.4324/9781315797991

Wheeler, R., 2014. Mining memories in a rural community: Landscape, temporality and place identity. *Journal of Rural Studies*, 36, 22–32. https://doi.org/10.1016/j.jrurstud.2014.06.005

Wicke, C., Berger, S., Golombek, J., 2018. *Industrial Heritage and Regional Identities*, Routledge, 1st edition.

Wikipedia, 2022. Ancient Greece. Accessed date: 18/02/2022. Available via: <https://tr.wikipedia.org/wiki/Antik_Yunanistan>

Wu, T., Xie, P.F., Tsai, M., 2015. Perceptions of attractiveness for salt heritage tourism: A tourist perspective. *Tourism Management*, 51, 201–209. https://doi.org/10.1016/j.tourman.2015.05.026

Xie, P.F., 2006. Developing industrial heritage tourism: A case study of the proposed jeep museum in Toledo, Ohio. *Tourism Management*, 27 (6), 1321–1330. https://doi.org/10.1016/j.tourman.2005.06.010

Xie, P.F., 2015. Industrial heritage tourism. Volume 43 in the series. *Tourism and Cultural Change*, Channel View Publications. https://doi.org/10.21832/9781845415143

Yang, X.S., 2017. Industrial heritage tourism development and city image reconstruction in Chinese Traditional Industrial Cities: A web content analysis. *Journal of Heritage Tourism*, 12 (3), 267–280. https://doi.org/10.1080/1743873X.2016.1236800

118   References

Yaşar, S., 2018. Berchtesgaden Tuz Madeni Müzesi (Almanya). (Berchtesgaden Salt Mine Museum (Germany)). *Mining Bulletin,* 127, 44–48. Available via: <www. maden.org.tr/resimler/ekler/565f268ed7f4d0a_ek.pdf>

Yaşar, S., 2019. Seegrotte Yeraltı Gölü ve Tarihi Jips Madeni (Seegrotte Underground Lake and Historic Gypsum Mine). *Mining Bulletin,* 131, 39–41. Available via: <www.maden.org.tr/resimler/ekler/18c797b65f5f909_ek.pdf>

Yaşar, S., 2020. İnsanoğlunun ilk Sistematik Sert Kaya Kazı Yöntemi: Ateş Kurma (Mankind's first systematic hard-rock excavation method: The fire-setting). *MT Bilimsel, Journal of Underground Resources,* 18, 67–78. Available via: <https://dergipark.org.tr/en/pub/mtb/issue/55984/767769>

Yazıcı Gökmen, E., Gülersoy, N.Z., 2018. Spatial planning as a tool for effective nature conservation: A conceptual framework for Turkey's spatial planning system. *Journal of Landscape Ecology,* 11 (1), 73–98. https://doi.org/10.2478/jlecol-2018–0002

Yıldız, T.D., 2020. Evaluation of forestland use in mining operation activities in Turkey in terms of sustainable natural resources. *Land Use Policy,* 96, 104638. https://doi. org/10.1016/j.landusepol.2020.104638

Yıldız, T.D., 2022. Türkiye'ye örnek bir endüstriyel miras: Lavrion Teknoloji ve Kültür Parkı. *IKSAD Publishing House, first edition,* ISBN: 978-625-8405-91-0, Available via: <https://iksadyayinevi.com/home/turkiyeye-ornek-bir-endustriyelmiras-lavrion-teknoloji-ve-kultur-parki/>

Yıldız, T.D., Samsunlu, A., Kural, O., 2016. Urban development and mining in Istanbul – Ağaçli Coal Field and its rehabilitation. (SWEMP 2016) Proceeding of 16th International Symposium on Environmental Issues and Waste Management in Energy and Mineral Production, 5–7 October 2016, Istanbul, Volume 29, pp. 1–11. Available via: <www.researchgate.net/publication/329962585_Urban_Development_and_Mining_in_Istanbul_-_Agacli_Coal_Field_and_Its_Rehabilitation>

Yılmaz, H., 2014. Eskişehir'in Biricik Destinasyon Önerisi Endüstriyel Miras (The unique destination proposition of Eskisehir: Industrial heritage). *Dokuz Eylül University Faculty of Economics and Administrative Sciences Journal,* 29 (2), 205–225. Available via: <https://dergipark.org.tr/tr/pub/deuiibfd/issue/22718/242470>

YMGV, 2015a. Travel to the past in the coal mine: Tour-Ed Mining Museum. *Turkish Mining Development Foundation (YMGV), Sector Mining Journal,* 56, 56–57. Available via: <https://cdn.ymgv.org.tr/cdn/uploads/urunler/Pdf/f9aa09ae-d27e-40f4-acf1-0633731377f1.pdf>56-57>

YMGV, 2015b. A mining museum is established in Zonguldak (News from Turkey). Turkish Mining Development Foundation (YMGV). *Sector Mining Journal,* 57, 14. Available via: <https://cdn.ymgv.org.tr/cdn/uploads/urunler/Pdf/9be23951-ae6f-4bdd-9861-60dea9cc1baa.pdf>

YMGV, 2016. Britannia Mines Museum. *Country Mining Development Foundation (YMGV), Mining Sector Magazine,* 59, 40–41. Available via: <https://cdn.ymgv.org. tr/cdn/ uploads/urunler/Pdf/bba553da-2b2d-4ed5–886b-18679ab1bf35.pdf>

Zouros, N., 2004. The European Geoparks Network. Geological heritage protection and local development. *Episodes,* 27, 165–171. https://doi.org/10.18814/epiiugs/2004/ v27i3/002 Quoted from (Voudouris et al., 2021).

Biography

Name-Surname: Taşkın Deniz Yıldız
Date/Place of Birth: 1984 / Istanbul-Turkey
E-mail: tdyildiz@atu.edu.tr

Articles Published in SCI, SSCI, SCI Expanded Index

Yıldız, T.D., Coşkun, N.D., Uz, V., İssi, A., Uz, B., 2023. Geological, petrographical, mineralogical, geochemical and gemological features of Malatya rubies, which are among the ruby deposits of the world. *Gospodarka Surowcami Mineralnymi: Mineral Resources Management,* 39 (4).

Yıldız, T.D., 2023. How can shares be increased for indigenous peoples in state rights paid by mining companies? An education incentive through direct contribution to the peoples. *Resources Policy,* 85 (A), 103948. https://doi.org/10.1016/j.resourpol.2023.103948

Yıldız, T.D., 2023. Changes in the salaries of mining engineers as they obtain managerial and OHS specialist positions in Turkey: By what criteria can salaries be increased? *Resources Policy,* 84, 103772. https://doi.org/10.1016/j.resourpol.2023.103772

Yıldız, T.D., 2022. Considering the recent increase in license fees in Turkey, how can the negative effect of the fees on the mining operating costs be reduced? *Resources Policy,* 77, 102660. https://doi.org/10.1016/j.resourpol.2022.102660

Yıldız, T.D., 2022. Supervisor fund expectation for the guarantee of salaries in the presence of the effect of permanent supervisor salaries on mining operating costs in Turkey. *Resources Policy,* 77, 102640. https://doi.org/10.1016/j.resourpol.2022.102640

Yıldız, T.D., 2022. How can the state rights be calculated by considering a high share of state right in mining operating costs in Turkey? *Resources Policy,* 75, 102509. https://doi.org/10.1016/j.resourpol.2021.102509

120 *Biography*

Yıldız, T.D., 2022. Are the compensations given to mining enterprises due to the overlapping with other investments in Turkey enough? Expectations of compensation for profit deprivation. *Resources Policy*, 75, 102507. https://doi.org/10.1016/j.resourpol.2021.102507

Yıldız, T.D., 2021. Loss of profits occurring due to the halting of mining operations arising from occupational accidents or reasons related to legislation. *Gospodarka Surowcami Mineralnymi: Mineral Resources Management*, 37 (4), 153–176. https://doi.org/10.24425/gsm.2021.139739

Yıldız, T.D., 2021. Overlapping of natural stone mining field with high-speed train project in Turkey: Was the economic public benefit evaluation made sufficiently? *Resources Policy*, 74, 102241. https://doi.org/10.1016/j.resourpol.2021.102241

Yıldız, T.D., 2021. Overlapping of mine sites and highway route in Turkey: Evaluation in terms of mining land use criteria and land use planning. *Land Use Policy*, 106, 105444. https://doi.org/10.1016/j.landusepol.2021.105444

Yıldız, T.D., 2021. How can the effects of EIA procedures and legislation foreseen for the mining operation activities to mining change positively in Turkey? *Resources Policy*, 72, 102018. https://doi.org/10.1016/j.resourpol.2021.102018

Yıldız, T.D., 2021. Possible effects of mining zone regulation on mining companies in Turkey & Evaluation of mining companies. *Resources Policy*, 71, 102011. https://doi.org/10.1016/j.resourpol.2021.102011

Yıldız, T.D., 2020. Evaluation of forestland use in mining operation activities in Turkey in terms of sustainable natural resources. *Land Use Policy*, 96, 104638. https://doi.org/10.1016/j.landusepol.2020.104638

Yıldız, T.D., 2020. Effects of the private land acquisition process and costs on mining enterprises before mining operation activities in Turkey. *Land Use Policy*, 97, 104784. https://doi.org/10.1016/j.landusepol.2020.104784

Yıldız, T.D., 2020. Forest fees paid to permit mining extractive operations on Turkey's forestlands & the ratio to investments. Gospodarka Surowcami Mineralnymi: *Mineral Resources Management*, 36 (3), 29–58. https://doi.org/10.24425/gsm.2020.133935

Yıldız, T.D., 2020. The impacts of EIA procedure on the mining sector in the permit process of mining operating activities & Turkey analysis. *Resources Policy*, 67, 101681. https://doi.org/10.1016/j.resourpol.2020.101681

Yıldız, T.D., 2020. Waste management costs (WMC) of mining companies in Turkey: Can waste recovery help meeting these costs? *Resources Policy*, 68, 101706. https://doi.org/10.1016/j.resourpol.2020.101706

Yıldız, T.D., Kural, O., 2020. The effects of the mining operation activities permit process on the mining sector in Turkey. *Resources Policy*, 69, 101868. https://doi.org/10.1016/j.resourpol.2020.101868

Articles Published in Other Indexed Journals

Yıldız, T.D., 2022. Required operating license before operation permit to be able to perform mining operation activities in Turkey. *Dokuz Eylul University the Journal of Graduate School of Social Sciences*, 24 (1), 119–146. https://doi.org/10.16953/deusosbil.680561

Yıldız, T.D., Gültekin, A.H., Özdamar, Ş., 2022. Gerede-Kavacık Bölgesi bazaltlarının endüstriyel kullanımı açısından uygunluğunun belirlenmesi (Determination of suitability for industrial use of basalts in the Gerede-Kavacık Region). *Journal of Engineering Sciences and Design*, 10 (1), 7–27. https://doi.org/10.21923/jesd.943867

Yıldız, T.D., Topaloğlu, M., 2021. Bringing mining rights as capitals to the trading companies from the perspective of valuations of licenses. *Dokuz Eylul University the Journal of Graduate School of Social Sciences*, 23 (3), 1123–1149. https://doi.org/10.16953/deusosbil.729746

Yıldız, T.D., Kural, O., Aslan, Z., 2021. To what extent is it possible to conduct mining activities in agricultural areas and especially olive groves: Solution expectations of enterprises. *The International Journal of Economic and Social Research*, 17 (1), 183–208. Available via: <https://dergipark.org.tr/tr/pub/esad/issue/62547/911629>

Yıldız, T.D., Maral, M., 2020. Supervision in the development of Turkish mining legislation. *Dokuz Eylul University the Journal of Graduate School of Social Sciences*, 22 (4), 1637–1677. https://doi.org/10.16953/deusosbil.767141

Yıldız, T.D., 2020. İşyeri açma ve çalışma ruhsatının (GSM) mevzuat ve madencilik sektörü açısından değerlendirilmesi: GSM ruhsatı ÇED kapsamına alınabilir mi? (Evaluation of business license and work permit (GSM) in terms of legislation and mining sector: Can GSM license be included in the scope of EIA?). *Journal of Cukurova University Faculty of Economics and Administrative Sciences*, 24 (2), 145–169. Available via: <https://dergipark.org.tr/tr/pub/cuiibfd/issue/60002/795152>

Yıldız, T.D., 2020. Recommendations for authorized administration organization in the mining operation permit process in Turkey. *Trakya University Journal of Social Science*, 22 (1), 117–143. https://doi.org/10.26468/trakyasobed.533814

Yıldız, T.D., Uz, B., Ülgen, S., Uz, V., Coşkun, N.H., Uçar, A., Kayıkçı, S., 2020. Bursa – Akçapınar – Kazanpınar civarında kireçtaşı kökenli mermer oluşumlarının etüt ve değerlendirilmesi (Survey and evaluation of limestone origin marble formations around Bursa – Akçapınar – Kazanpınar). *Journal of Engineering Science of Adıyaman University*, 7 (13), 56–74. Available via: <https://dergipark.org.tr/tr/pub/adyumbd/issue/58754/733823>

Yıldız, T.D., 2020. Forest costs paid by enterprises during investment period to carry out mining operations in forestlands. *Journal of Engineering Science of Adıyaman University*, 7 (12), 24–33. Available via: <https://dergipark.org.tr/tr/pub/adyumbd/issue/54711/704953>

Uz, V., Uz, B., İssi, A., Coşkun, N.D., Yıldız, T.D., 2018. The Annealing of Corundum (Ruby) in Nitrogen (N_2) Air. *El-Cezeri Journal of Science and Engineering*, 5 (3), 875–881. https://doi.org/10.31202/ecjse.436515

Yıldız, T.D., 2019. The share of required costs in investment amounts for mining operating activities in pasture lands in Turkey. *Journal of Engineering Science of Adıyaman University*, 6 (10), 23–31. Available via: <https://dergipark.org.tr/tr/pub/adyumbd/issue/45775/555222>

Reviewed Congress/Symposium Publications in Proceedings

Bilgi, S., Özdamar, Ş., Yıldız, T.D., Uz, B., Kalkan, Y., 2020. Zemin kayacının hafriyatı sonucu alınan malzemenin jeolojik – petrografik özellikleri ve miktarının yerinde jeodezik ölçümler ile belirlenmesi (Geological – petrographical properties of the material taken by excavation of ground rock and determination of its quantity by geodetic measurements). 3rd International Congress of Academic Research, July 20–22,

122 Biography

pp. 427–445. Available via: <https://kongre.akademikiletisim.com/files/icar3/icar3_tam_metin.pdf>

Bulut, M., Yıldız, T.D., 2023. Agrega tesislerinde bulunan filtre pres ünitelerindeki atık çamurlarının değerlendirilmesi (Evaluation of waste sludge in filter press units in aggregate facilities). 11th Euroasia International Congress on Scientific Research and Recent Trends-11, June 22–23, Goreme Municipality – Cappadocia, Nevşehir, Proceedings Book, pp. 29–31. Available via: <https://f749d2de-b9db-4cec-b6d0-ee07a424fefb.filesusr.com/ugd/262ebf_48fdd5bd04394475bb1c4648f7a6eb58.pdf>

Uz, B., Çoban, F., Yıldız, T.D., Ersoy, A., Uz, V., Kural, O., Maral, M., 2017. Muğla Yatağan – Yeniköy – Berber Mahallesi civarının kalsit açısından etüt ve değerlendirmesi (Measurement and evaluation of Muğla – Yatağan – Yeniköy – Berber Quarter in terms of calcite). 4th International Stone Congress, March 20–25, pp. 533–546. Available via: <www.researchgate.net/publication/329962410_Mugla_Yatagan_-_Yenikoy_-_Berber_Mahallesi_Civarinin_Kalsit_Acisindan_Etut_ve_Degerlendirmesi>

Uz, B., Özdamar, Ş., Gültekin, A.H., Yıldız, T.D., Uz, V., Ersoy, A., Yünsel, T.Y., Kural, O. and Maral, M., 2018. Mining and Geological Investigation of the Coal Field in Tekirdağ-Malkara & Reserve Account. INERMA (2nd International Energy Raw Materials and Energy Summit), September, 27–30, Turkish Mining Development Foundation Publications, pp. 51–58. Available via: <www.researchgate.net/publication/329962605_Mining_and_Geological_Investigation_of_the_Coal_Field_in_Tekirdag-Malkara_Reserve_Account>

Uz, B., Yıldız, T.D., 2017. Milas Akbük Ege bordo mermerlerinin etüt ve değerlendirilmesi (Measurement and evaluation of Milas Akbük Aegean Claret-Red Marbles). Turkey 9th International Congress and Exhibition of Natural Stones and Marble, December 13–15, pp. 545–556.

Uz, B., Yıldız, T.D., 2017. Bursa – Doğanalan – Körekem civarında mermer oluşturan kireçtaşlarının etüt ve değerlendirilmesi (Measurement and evaluation of marble-forming limestones around Bursa – Doğanalan – Körekem). Turkey 9th International Congress and Exhibition of Natural Stones and Marble, December 13–15, pp. 557–569.

Uz, B., Yıldız, T.D., Ersoy, A., Uz, V., Maral, M., Işık, L., 2017. Gebze – Mudarlı kireçtaşlarının doğal taş açısından etüt ve değerlendirmesi (Measurement and evaluation of the limestones with regard to natural stone-marble around Gebze/Mudarlı). 4th International Stone Congress, March 20–25, pp. 509–532. Available via: <www.researchgate.net/publication/329962506_Gebze_-_Mudarli_Kirectaslarinin_Dogal_Tas_Acisindan_Etut_ve_Degerlendirmesi>

Uz, B., Yıldız, T.D., Gültekin, A.H., Esenli, F., Özdamar, Ş. ve Kural, O., 2017. Bir kalker sahasında yapılacak patlatmanın arkeolojik yapılara ve SİT alanlarına etkisi (The effect of blasting on a limestone field to the archaeological sites and protected areas). 9th International Drilling Blasting Symposium, December 1–2, pp. 27–38.

Yıldız, T.D., 2013. Türk maden mevzuatı evriminde madenlerin mülkiyeti ve rejimi (The ownership of regime of mines in the evolution of Turkish mining legislation). Proceedings of the 23rd International Mining Congress and Exhibition of Turkey, April 16–19, pp. 1959–1980.

Yıldız, T.D., 2013. 3213 sayılı Maden Kanunu öncesinde ve sonrasında maden arama faaliyetlerindeki değişikliklerin incelenmesi (Analysis of the changes in mining exploration activities before and after Mining Law No. 3213). Proceedings of the 23rd International Mining Congress and Exhibition of Turkey, April 16–19, pp. 1981–1994.

Yıldız, T.D., 2017. Türkiye'de havalandırma konusundaki mevzuat hükümlerinin iş sağlığı ve güvenliği açısından değerlendirilmesi (Evaluation of legislation provisions on ventilation in Turkey in terms of occupational health and safety). 25th International Mining Congress and Exhibition, April 11–14, pp. 270–282. Available via: <www.researchgate.net/publication/329962730_Turkiye'de_Havalandirma_Konusundaki_Mevzuat_Hukumlerinin_Is_Sagligi_ve_Guvenligi_Acisindan_Degerlendirilmesi>

Yıldız, T.D., 2020. Madencilik sektörünün acil olarak çözülmesi gereken sorunları (Problems of the mining sector that need to be solved urgently). 10th International Science and Technology Conference (ISTEC 2020), September 3–4, Nicosia, Turkish Republic of Northern Cyprus, Abstract Book, pp. 96–97. Available via: <http://iste-c.net/publication_folder/inte/inte-istec-iticm-ietc-iwsc-2020_abstract_book.pdf>

Yıldız, T.D., 2021. Maden işletme faaliyetleri ile diğer yatırım faaliyetlerinin çakışmalarında mağduriyet yaşanmaması için neler yapılabilir? (For enterprises not to suffer, what can be done to avoid overlapping of mining operations and other investment activities?) 6th International Stone Congress, August 24–27, 2021.

Yıldız, T.D., Güner, M.O., Kural, O., 2017. Türkiye'de Maden Atıkları Yönetmeliği'nin madencilik sektörüne etkileri (The effects of the mineral waste regulation in Turkey on the mining sector). 25th International Mining Congress and Exhibition, April 11–14, pp. 457–472. Available via: <www.researchgate.net/publication/329962843_Turkiye'de_Maden_Atiklari_Yonetmeligi'nin_Madencilik_Sektorune_Etkileri>

Yıldız, T.D., Haner, B., 2017. Yeraltı madenciliğinde ocağa verilmesi gereken hava miktarı konusunda Türk mevzuatı hükümlerinin değerlendirilmesi (Evaluation of Turkish legislation provisions on air quantity to be submitted to underground minerals and ventilation ways). International Symposium on Occupational Health and Safety in Mining'2017, November 02–03, pp. 47–65. Available via: <www.maden.org.tr/resimler/ekler/8d3eefe2cdb3628_ek.pdf>

Yıldız, T.D., Kural, O., 2019. Maden işletme faaliyetleri ile hazinenin özel mülkiyet arazilerinin çakışması halinde istenebilecek bedeller & Mevzuat değerlendirmesi (Costs to be required in the conflict of mining operation activities and private property land of state's treasury & Evaluation of legislation). 1st International Congress of Academic Research (September16–18, 2019), Conference Abstracts, pp. 201–202. Available via: <https://kongre.akademiletisim.com/files/icar2019/icar_tam_metin.pdf>

Yıldız, T.D., Kural, O., Aslan, Z., 2019. Türkiye'de orman alanlarında maden işletme faaliyetleri yapılabilmesi için gerekli izinler konusunda yaşanan sorunlar ve çözüm önerileri (Problems and solutions in relation to permits required to be perform mining operation activities in forest lands in Turkey). 1st International Şişli Science Congress, October 24–25, pp. 159–160. Available via: <www.sisli.edu.tr/uisbk/2019/tr/wp-content/uploads/2019/09/issc-abstractsbook.pdf>

Yıldız, T.D., Kural, O., Çatan, B.E., 2019. Mera alanlarında maden işletme faaliyetleri yapılabilmesi için öngörülen izin süreci (Permission process for perform mining operation activities in pasture areas). International 30rd August Scientific Research Symposium, August 28–31, Full Text Book of Applied Sciences, Iksad Publications, pp. 272–280. Available via: <https://cb9043b8-89d3-4909-bcf8-ff6b2f5b97a0.filesusr.com/ugd/614b1f_5204de669e864dafbeebed6a90473405.pdf>

124 Biography

Yıldız, T.D., Maral, M., 2022. Evaluation of workers from the perspective of permanent supervisors in providing occupational health and safety in mining enterprises. 3rd Global Research on Multidisciplinary Sciences, July 4–7, Abstracts Book, ISBN: 978-625-7148-43-6, pp. 43–44, Istanbul, Turkey. Available via: <https://drive. google.com/file/d/1IO8Tl_BRUQUZWuNpA_ILPtyqhxEla0G2/view>

Yıldız, T.D., Samsunlu, A., Kural, O., 2016. Urban development and mining in Istanbul – Ağaçli Coal Field and its rehabilitation. (SWEMP 2016) Proceeding of 16th International Symposium on Environmental Issues and Waste Management in Energy and Mineral Production, October 5–7, Volume 29, pp. 1–11. Proceedings Citation Index-Science is indexed by Thomson Reuters CPCI-S, WOS No: 000391287200029. Available via: <www.researchgate.net/publication/329962585_Urban_Development_and_Mining_in_Istanbul_-_Agacli_Coal_Field_and_Its_Rehabilitation>

Books and Book Chapters

Maral, M., Yıldız, T.D., 2020. *Nezaretçilerin gözünden maden işletmelerinin değerlendirilmesi (Evaluation of Mining Operations from the Perspective of the Supervisors)*, O. Kural (editor). IKSAD Publishing House, 1st edition, p. 162. Available via: <https://iksadyayinevi.com/product/nezaretcilerin-gozunden-maden-isletmelerinin-degerlendirilmesi/>

Yıldız, T.D., 2020. *İşletme izin sürecinin madencilik sektörüne etkileri (Effects of operation permission processes on the mining sector)*, O. Kural and Z. Aslan (editors), IKSAD Publishing House, 1st edition, p. 394. Available via: <https://iksadyayinevi. com/wp-content/uploads/2020/09/SLETME-IZIN-SURECININ-MADENCILIK-SEKTORUNE-ETKILERI.pdf>

Yıldız, T.D., 2022. *Türkiye'ye örnek bir endüstriyel miras: Lavrion Teknoloji ve Kültür Parkı*, IKSAD Publishing House, 1st edition. Available via: <https://iksadyayinevi. com/wp-content/uploads/2022/03/TURKIYEYE-ORNEK-BIR-ENDUSTRIYEL-MIRAS-LAVRION-TEKNOLOJI-VE-KULTUR-PARKI.pdf>

Yıldız, T.D., Kural, O., Aslan, Z., 2020. Türkiye'de orman alanlarında maden işletme faaliyetleri yapılabilmesi için gerekli izinler konusunda yaşanan sorunlar ve çözüm önerileri (Problems and solutions in relation to permits required to be perform mining operation activities in forest lands in Turkey). Chapter 2 in book. *Academic Studies in the Field of Science and Mathematics*, Gece Bookstore Publishing, pp. 23–46. Available via: <www.gecekitapligi.com/Webkontrol/uploads/Fck/FenBilimlerive-Matematik.pdf>

Memberships

- Founding member of Association International Network Acclimating post-mining territories (October 2022-)
- Professional member of The Association of Geoscience, Mining and Metallurgy Professionals (YERMAM), available via: <http://yermam.org.tr/uyeler/uyeler> (March 2021-)
- Member of Professional Development Association of Mining Engineers (MMMGD), Available via: <www.mmmgd.org.tr/en/>, (October 2021-)

- Member of European Association of Labour Economists (September 2022)
- Member of Chamber of Mining Engineers of Turkey (2008–)

Awards

- Leadership Awards: International Research Awards on Leadership and Management under the category of Outstanding Scientist Award, by ScienceFather, India, Date: 10/07/2023. Available via: <https://leadership-conferences.scifat.com/awards-winners/>
- Number of Published Articles in WOS Category, University Award, 3rd academician rank between 2019–2022, Awarded by: Adana Alparslan Türkeş Science and Technology University, 25/10/2022.
- 2023 TUBITAK Academic Incentive Award (15000 Turkish Liras (TL))
- 2022 TUBITAK Academic Incentive Award (10000 TL)
- In International Research Awards on New Science Inventions under the category of Best Researcher Award, by ScienceFather, Date: 16/11/2022. Available via: <https://new-science-inventions.sciencefather.com/digital-health-awards-winners/?gv_id=&gv_search=Turkey&mode=any>
- In International Research Awards on Science Health and Engineering under the category of Best Researcher Award. SHEN 2021 Awards, by Science-Father, Approved and Registered by Ministry of Corporate Affairs (MCA), Govt. of India. Date: 18/05/2021. Available via: <https://shen.sciencefather.com/view/award-winners/?gv_id=&gv_search=Turkey&mode=any> youTube Channel video link: <https://youtu.be/X8y37E06aNU>
- 2021 TUBITAK Academic Incentive Award (5000 TL)
- 2020 TUBITAK Academic Incentive Award (5000 TL)

Index

abandoned buildings 76; destruction, threat 10; restoration 90
abandoned coal mine (Gunkanjima Island) 20
abandoned coal mines, evaluation 20
abandoned mine area: natural setting 72; tourism, possibility 21
abandoned mine-metallurgical plants, purchase 2
abandoned mine-metallurgy plants, restoration 5
abandoned mines: active mining 22; presence 51
abandoned mine sites 90; air quality, characterization (determination) 64; determination 4–5; mine/ining (production) heritage, usage 13; natural setting 72–73; organization 20; training activities, conducting 91–92
abandoned mining plants, reuse/restoration 51–63
academic internship (Atlas system) 70
acid mine drainage 65
acid-producing resources, sulfide waste generation 64–65
acid-producing wastes 64–65
Acropolis temples, construction 33–34
activity area, concordance 22
adits 6, 74, 77–79; expansion 30, 86; intruders, entry 87; opening 1, 33, 78; operation 34; presence 80–81; re-opening 1, 86; restoration 87; revealing 83–84; underground adits, entrances 3; unlocking 87; widening, drainage procedures (usage) 83; see also ancient adits

Agios Konstantinos, Mineral and Mining Museum 77
Alexander the Great 28
Almaden-Spain Mercury Mine 23
ancient adits: closure 87; length 78; mapping 87
ancient age: adits, opening 78; clothes, presence 72; iron, production 80; lead/silver mine site production 2; minerals, production 80; mining/metallurgical activities 30
Ancient Age, social classes/social unrest (growth/increase) 28
ancient Greece, Lavrion deposits (impact) 33
Ancient Lavrion: data 81–82; galenite melting method 83
ancient Lavrion, data 81–82
ancient mine: center (Thorikon) 33; plants 84; regions 88; shafts 6; sites, world heritage list 20; tunnels 69
archaeology, evaluation 51
archeological ruins, examination/classification 54–55
architecture: evaluation 51; FMC buildings 42; industrial architecture, examples 52–53; industrial heritage concept, relationship 8; industrial plant 55; monument 2; neoclassical architecture, example 60; repurposing, usage 10; samples 13; tangibles values, relationship 8; unique work 54
arsenic: compounds, waste 66; levels, soil analysis 67
artists, support 46

asbestos, pollution 4
Athenian Democracy, Golden Age 29–30
Athenian state, funding 29
Athens, Golden Age 29
Athens International Airport 72
Attica 43; business/culture center 44;
    dense population 64; Lavreotiki
    Geopark 76; Lavreotiki
    region location 51, 74;
    Lavrion mines 88; LTCP 69;
    LTCP technological park 44;
    microclimate 76; pinewoods 76;
    sea tourism area 62; territory
    40–41
Attica region: financing 38, 40;
    landscape 51–52; LTCP
    innovation center 45;
    partnership 70; secondary
    self-governing unit 41

Bank of Constantinople 31
barrage, creation 66
Belavilas, N. 54
Berchtesgaden Salt Mine Museum 23
Birds Diretive 79/409/EE 41
Blegny Coal Mine Museum 23
Bronze Age 14, 27–28; copper
    production 80; Late Bronze
    Age, copper resource 79–80;
    Lavrion mines, impact 29;
    mineral resources, extraction
    79–80
buildings: block, cultural heritage 58;
    decontamination 40, 46–47;
    historical-cultural value 10; *see
    also* Lavrion buildings
building stocks: abundance 7;
    importance 11; usage 10
business administration/management,
    methods 45
business development center (BIC) 70;
    establishment 40–41

cadmium: levels, soil analysis 67;
    pollution 4
Canadian Britannia Mine Museum 23
Çankırı Salt Cave Museum 24
carbonated zinc, accumulation 81
Castellano furnaces 31, 84
cerussite: ores, heaviness 82; production/
    processing 80
cerussite-originating silver carbonate,
    accumulation 81

chalcopyrite 3
children, museum education 22
Church of Agia Paraskevi *61*, 74
city preservation projects 11
classical age 1, 28
Classical Age, buildings 32
Classical age, Lavreotiki silver/lead
    supplier 74
Classical Greek Period, silver
    production 82
clay layers, HDPE liners (usage) 65, *65*
Clemence deposit 3
Coal Mine Museum *see* Blegny Coal
    Mine Museum; Old Coal Mine
    Musem; Tour-Ed Coal Mine
    Museum
coin (Lavrion silver composition) *30*
collective memory: local community,
    relationship 59; preservation/
    building 10; revival 38; visual/
    spatial traces 11
Comarca Minera, Global Geopark
    (recognition) 20
commercial activities, creation/
    development 44
Compagnie Francaise, founding 31
Company for the Utilization and
    Management of the Property
    of the NTUA (CUMP-NTUA):
    formation 39; operation 49
Convention on International Trade in
    Endangered Species of Wild
    Fauna and Flora (CITES) 64
cooperation center, establishment 22
copper: extraction 79–80; levels, soil
    analysis 67; production 80;
    supply 29
Copper Age 14, 27
Cornwall Mining District 23
cultural activities, hosting/organizing/
    producing 46
cultural context, highlighting 44
cultural development 12, 41; technology
    basis 40
cultural heritage: abandoned coal mine
    evaluation 20; concept, handling
    25; documentation/display
    58; evaluation 14; industrial
    heritage, relationship 8;
    landscape, relationship 51–52;
    Lavreotiki region acceptability
    85–86; Lavrion deposits,
    impact 33; preservation 16–17;

relationships 15; tourism 19;
  value 4; value, industrial ruins
  (involvement) 7
cultural identity: formation 10; revival
  21; strengthening 21
cultural tourism: experiences 19–20;
  importance, increase 15;
  Lavrion center 76; mining
  tourism, relationship 20
Cyclades 62

"dark age" 28
Dark Ages 28
decontamination 40; issues 46–47;
  process, emergence 39
deindustrialization (de-industrialization)
  9, 35, 37
"Designation of Protected Mountainous
  Zones of Lavreotiki Peninsula"
  (Gazette No 121/D) 42
digging activities 66
Documentation and Conservation
  of Buildings, Sites,
  Neighbourhoods of the Modern
  Movement (DOCOMOMO) 9
dressing plants 83–84

economic activities: direction 16–17;
  Lavrion buildings presentation
  62, 63
economic crisis: cycle, Lavrion
  (relationship) 35, 37; emergence
  34–35; Greece 90; impact 2,
  39–40; socio-economic crisis
  37–38
economic problems 84; appearance 2;
  impact 29
education activities, LTCP offering
  69–70
educational activities: hosting/
  organizing/producing 45;
  promotion 86
electric power plan (LTCP) 52
employment, Lavrion mines (effects)
  33–35
engineering values, tangibles values
  (relationship) 8
enterprises, attention (attraction) 44
environmental conditions, application/
  improvement 65, 68
environmental criteria, usage 64–65
Environmental Education Centre of
  Lavreotiki (EECL) 49, 72

Environmental Education Centre
  of Lavreotiki, educational
  institutions 49
environmental hazard, control 64
environmental measurements:
  consideration/establishment 49,
  66; laboratory 50
environmental pollution: problem 65;
  risk 64
environmental projects (Lavrion) 63–69
environmental protection: application 65;
  project 64
ethnography, evaluation 51
EU Framework Programme,
  implementation 39
European Ecological Network NATURA
  2000 85
European Geoparks Network 20
European Landscape Convention 51
European Union (EU) funding 23
Evangelistria, construction 74
Evangelistria Orthodox Church 60
excavations, revelations 83–84

Falun Sweden Great Copper
  Mountain 23
Fe-Mn ores 80, 81
flat-type mineral processing 83
flotation building: exposure 58;
  images/photos 59, 59; three-
  dimensional perspectiveless
  view 58
Flotation plant ("La Flotation"):
  post-restoration 59; restoration/
  reuse 58–59; three-dimensional/
  perspectiveless view 58
foreign investment, attraction 45
"For the Protection of the World
  Cultural and Natural Heritage"
  (UNESCO Convention) 41
French Mining Company (FMC):
  activity 37; archives 56, 56,
  59, 87; closure 52; converted/
  restored buildings 54; former
  settlement, LTCP founding
  38; foundation 1; founding
  32; lead-zinc deposits 82;
  management/auxiliary buildings
  53; old buildings, reuse 2; ore
  extraction 82; role (Lavrion
  development) 52; site 60, 72–73
French Mining Company (FMC)
  buildings 55; conversion 43;

settlement/destruction, planning
37–38; transformation 53–54
French Mining Company (FMC) plants
55; rescue/restoration 38
French Ore Loading Port 74
funding sources 39
furnaces, equipping 30–31

galena 3, 30, 82
galenite: melting method 83; ores,
heaviness 82; presence 81;
production 80
geodiversity (geo-diversity) 14; concept
71; creation 15; management
17; natural capital, non-living
component 15; preservation 15;
shaping 16
geography, evaluation 51
geoheritage 5, 13, 14–17; areas 90;
assets 16; characteristics 6, 74;
evaluation 17; preservation 23;
sites 14, 91; tourism 22; values 14
geological characteristics 76–77
geological heritage: preservation
16–17, 20; protection 86; site,
Lavreotiki region acceptability
85–86; sites 17; term, usage 15
geological processes 15, 16, 20
geological tourism (geotourism) 15;
activities 17; areas, accessibility
16; development dynamics,
indication 21; evaluation
14; interest, increase 20;
mining tourism, relationship
20; objectives 85; planning
20; project, success 20–21;
sustainability 21
geology, evaluation 51
geopark 90; establishment 85; plans,
values demonstration 16–17;
project 6, 85–87; protection 86;
qualification 3
Geopark: Lavreotiki Geopark 76; project,
preservation 85–86
geosites 15–16; Megali Cariera geosite
80; recognition 17
geotechnical conditions 69
geotourism *see* geological tourism
geotrails series 74–75
Global GeoPark Network 16
Global Geopark, recognition 20
Global Geoparks Network/Networks 16,
85–86

gold deposits, discovery 34
golden age 28
Golden Age (Athens) 29
Golden Age of Athenian Democracy
29–30
Golden Age of Pericles 1
Goulandris Natural History Museum
(Athens) 88
Greece: economic crisis 90; economic
problems, impact 29; history,
overview 27–29; immovables
42; real estate, UNESCO World
Heritage Sites tentative list **42**
Greek citizens, history/urban
conservation awareness 4
Greek colonies, politics control 28
Greek culture: ancient cultures, merger
28–29; development 90; spread
28–29
Greek government: mine-metallurgical
plants purchase 2; NTUA
appointment 38–39;
technological park
establishment 38
Greek history 90; explanation 5;
Hellenistic Age 28; initiation
27; shaping 4
Greek industrial activities 37
Greek Lavrion Metallurgical
Company 31
Greek Metallurgical Company 60, 62
Greek Mining Company 82

Handicraft and Industrial Education
Museum 47, 72
hazardous waste: containment 67; LTCP
storage *69*; storage *68*
heavy metal: concentrations 67; pollution
63–64
Hellenistic Age 28
heritage management 9; processes 12
heritage protection: failure 11;
impossibility 12; tourism
attractiveness development
strategies, conflict 22
high-density polyethylene (HDPE) liners
65, *65*
high-tech sectors, foreign investment
(attraction) 45
historical building, changes 58
historical heritage sites 21–22
historical identity, revival 38
historic buildings, upgrade 45

historic houses 72
historic mine sites 13, 20
history, evaluation 51
hydraulic-mechanical processing, metal
    production 81
hydro-mechanical mineral processing 84

Ilarion Roux et Compagnie 82
inclined shafts 74, 76
industrial activities: Greek industrial
    activities 37; hosting 42;
    housing 53; records 12
industrial activities (Lavrion) 62–63
industrial archaeology: FMC buildings
    42; monument 2; unique
    work 54
industrial architecture, examples 52
industrial archives, importance 12, 40, 55
industrial area, environmental issues 50
industrial buildings 60, 62; demotion/
    functional loss 7; description
    54–55; design 58; evaluation
    11; importance 10; protection/
    preservation 7; restoration
    63; reuse 14; threats 8;
    transformation 11–12
industrial character 52
industrial crisis 5, 37–38
industrial equipment: historical archive
    46; protection 52–53
industrial equipment (LTCP) 53
industrial esthetics 49
industrial heritage: abandoned coal mine
    evaluation 20; area 5, 45, 86,
    90; birth 7–9; conservation 12;
    cultural heritage, relationship
    8; landscape, relationship
    51; preservation/reuse 9–12;
    sites 7–9, 12–13, 42; sites,
    declarations 90–91; tourism
    17–21
industrial heritage, concept 7–9; evaluation
    8; reuse/preservation 7
industrial history 4, 5, 12, 37;
    understanding/interpreting 12
industrialization 34; city encounter
    10; deindustrialization
    (de-industrialization) 9, 35, 37;
    developing industrialization
    1; impact 8–9, 11; increase
    7–8; process 10–11; production
    increase, impact 7

industrial monuments 72; historic
    content-listed industrial
    monuments 42; museum
    establishment 23; preservation 18
industrial plant: architecture 55;
    importance 52
industrial production: decline 18;
    processes 8
industrial revolution 7, 13
Industrial Revolution 18
industrial ruins, involvement 7
industrial site 7–8, 12, 13
industrial zone 33, 58
information production organizations,
    enterprises (connection) 45
innovative cultural activities, support 46
innovative use plan: development/testing
    45; implementation 45
innovative work, complement/securing
    45
intangible heritage, recovery 45
International Athens Airport 62
Iron Age 14, 27
iron, production 80
isotope analysis 79

J.B. Serpieri Monument 74

Kamariza: FMC mining activities 86–87;
    mineral deposits, discovery 33;
    mineralogy museums, exhibits
    89; mining center/section/area
    31, 77, 80; ores, mining 84;
    region, carbonate-substituted
    ores (extraction) 3
Kamariza Railway 79; tunnel 32; usage
    78–79
Kamariza Series (Kamariza series) 78,
    81
Keratea, Plaka region 81
knapweed species (Centaurea laureotica)
    (Centaurea attica) 76
knowledge, acquisition/dissemination 45
Konofagos building, decontamination 40
Kordelas, A. 84
Kordellas, A. 30

land: agricultural land 51–52; evaluation
    57; forces 28; future 85;
    industrial land, transformation
    21; metal-producing land 29–30;
    polluted/degraded land 39;

rehabilitation work emergency risks, reduction 66–67; sustainable method 15
landfill, soils accumulation 66
landscape, importance 51
land use 42, 52; planning methods, development 11; types 13
Late Bronze Age, copper resource 80
Lavreotika Affair 32
Lavreotiki 43; discovery area 71; forested areas, protection 42; geological landscape 74; geology 71; landscape, change 52; Para-otokton series 81; preservation 41–42; protection 6, 42; riches 34; silver/lead supplier 74; Sounion National Park 64
Lavreotiki Geopark 39, 76, 86; integration/inclusion 85–86
Lavreotiki mines 80; FMC operation 60
Lavreotiki municipality 43, 70; population, decline 35
Lavreotiki region 50, 51, 66, 78; acceptability, protection 86; importance 3; location 51
Lavrion 35; abandoned mining plants, reuse/restoration 51–53; adits 77–79; boat docking 63; development, FMC role 52; economic crisis cycle 35, 37; education/activities 69–73; filming 73; founding 52; Geopark project, preservation 85–87; history 27; houses 57; indigenous people 35; industrial activities 62–63; landscape 51–53; landscaping 5–6; location 43, 75; metallurgical plant, machine shops 31; metallurgical plants/port 31; metallurgical plants/railway 31; metallurgy plants, mineral processing/ruins 82–85; mineral deposits, features 79–82; mineral deposits, importance (decline) 34; mining, history 29–32; mining-metallurgy plants, activities 33; morphology 75; movie scenes 73; municipality 85; plants/equipment, reuse

discussion/options 38; plants, layout 56; railroad, transport activities 79; railways 77–79; rehabilitation/environmental projects/activities 64–69; rehabilitation projects 64–67; restored buildings 60, 62; silver 30
Lavrion buildings 60, 62; economic activities 62–63, 63; reuse/restoration projects 53–55
Lavrion deposits: mineralization types 88–89; research 71
Lavrion Mineralogy and Archeology Museum 6, 89
Lavrion mines: abandonment 30; effects 33–35; employees 34; impact 5–6, 29–30, 33–35; importance 3–4, 86; old mine shafts 77–79; production 30, 33; workers, accommodation 77
Lavrion mine sites: geoheritage characteristics 74; industrial/mining heritage 74
Lavrion Mining and Metallurgical Museum (LMMM): construction/installation 55; historical museums 88; involvement 40; machine buildings 53; Machine Buildings housing 52–53
Lavrion museum *see* Mineralogy and Archeology Museum of Lavrion
Lavrion Port 39, 40, 62–63, 78
Lavrion Port Authority 40
Lavrion Technical and Cultural Park (LTCP): Attica technological park 44; backbone/circulation 55; birth 37–38; boundaries, underground area (creation) 66; business/enterprise types, union/development 45; communication system, design/implementation 44; cooperation, development 44; electric power plant 52; environmental measurements 50; facility design, Faculty of Architecture team management 55; field, drone image 57; financing 37; founding 38; functions/management/

finance 5; goals 44; industrial
equipment *53*; innovation
center 45; laboratories/
companies, research/educational
collaborations (mediation) 70;
landscape 51–53; landscaping
5–6; location 43; machine
shop *53*; management 2, 39,
68; master plan 55–58; master
plan, drawing 57; plants 49, 53,
55; post-restoration LTCP area
A buildings *47*; provisions 70;
region, building locations *48*;
region, institutions/companies
47; research projects *51*;
restoration plan 55–56; restored
area *46*; reusing suggestion
37–38; science/cultural park
function 57; storage *69*;
underground hazardous waste
storage *68*; underground
storage, construction works/
adit *67*; underground waste
repository, features 66–69
Lavrion Technical and Cultural
Park (LTCP) area: digging
activities 66; reuse/restoration/
rehabilitation, achievement 5
Lavrion Technical and Cultural Park
(LTCP) buildings: educational
activities *70*; three-dimensional
spatial/historical process,
evaluation 57
Lavrion Technical and Cultural Park
(LTCP) projects 37; activities 4,
37; aims 43–44; features 44–45;
location 43; management/
financing 38–41; scopes 43,
44–45
Lavrion Technical and Cultural Park
(LTCP) sites 67; aims 43–44;
evaluation 44; rehabilitation/
restoration 37
Law 360/76 42
Law 998/79 42
Law 1650/86 41
Law 2742/99 41
lead: carbonate-involving lead 80;
extraction 79–80; isotope
analysis 79; Lavreotiki supplier
74–75; levels, soil analysis 67;
low-grade tailings 82; mine sites
2; pollution 4; price, decline

34–35; recovery furnaces 84;
sheet, production 85; silver-lead
ore, processing 77; supply 29
lead ore: extraction 81; usage 1
lead-originated silver carbonate,
accumulation 81
lead-zinc deposits (FMC) 82
lead-zinc-iron-copper ores 80
local administrations, promotions
(making) 22
local community, collective memory 59
low-grade ores 31–32, 78, 81–82
LTCP Laboratory 65, 70

Macedonia: deposits 3; Greek culture,
spread 28–29; silver/gold
deposits 34
Macedonian Empire, establishment 28
Machine Building 40, 53
machine equipment, preservation 42
machine shops 84; housing 31;
metallurgical plant location *31*
magnetite ore deposit 81
Master Plan (master plan) 55–58
mechanical mineral processing plants 84
Megali Cariera geosite 80
melting pot, constant airflow (impact)
82–83
Metal Age 27
metallurgical activities 1, 13, 30, 37,
63–65, 84
Metallurgical Company of Lavrio 84
metallurgical furnaces, usage 3
metallurgical plants 89
metallurgical process 57
metallurgical wastes 64
metal production 81
mine: acid mine drainage 63–64, 65;
FMC mine workers, transport/
wagon usage 89; villages,
company center connection 52;
workers, habits/accommodation
13, 77
mine-metallurgical plants 2; Lavrion
location *31*
Mineral and Mining Museum (Agios
Konstantinos) 77
mineral deposits: activity 14; features 6;
importance, decline 34; richness
(Maroneia) 33
mineral exploration 4, 24, 87
mineral heritage sites 17
mineral museum 23–25

mineralogy 4; formation 71; information
87–88; Lavrion, importance
88; Lavrion Mineralogy
and Archeology Museum
6, 89; museum 2, 89;
three-dimensional museum 87
Mineralogy and Archeology Museum 74
Mineralogy and Archeology Museum of
Lavrion 88–89
mineral processing 4, 13, 82–85; activity,
wastes 64; focus 80; furnaces
74; plants 31, 53, 76, 78; ruins
6; sites 8; units 3, 30, 58
Mines du Camariza 31
mine shafts 3, 6, 74, 86; old mine shafts
77–79
mining activity 13–14; cessation 35;
mine site 78–79
Mining and Mineral Museum 23–25
mining heritage 13–14; preservation 23;
relationships 15; sites 4–5, 14,
21–22, 91–92; tourism 5, 17,
19–22
mining history 5, 13, 19, 22, 24;
examination 71; Lavrion 33, 77;
learning 89
mining industry 13; development 34;
heritage 13, 14
mining-metallurgical activities,
development 66
mining-metallurgy: activities 72; centers
32; plants, activities *33*; system
(Lavrion) 87
mining museum 5, 20, 23–25; building
25; foundation 14; list 23–24;
visitor attractions 19
mining sector 20
mining tailings 30, 31
mining tourism: development 19–20;
interest, limitation 19;
opening 20
ministries of culture and tourism 5
ministry of culture and tourism 17,
22, 91
ministry of energy and natural resources
17, 22
Minoan civilizations, existence 27–28
Minoan Period 29
mixed sulfide lead-zinc-iron-copper
ores 80
modern Lavrion, data 82
monument: concept 54; natural
monument 16

movie scenes (Lavrion) *73*
municipality: Lavrion 85
municipality (Lavreotiki) 35, 43, 70
municipality building, neoclassicism
60–61
Municipal Public Benefit Agency of
Lavreotiki 85
Museum of Mineralogy and Petrology
(University of Athens) 88
Museum of Mineralogy & Paleontology
Stamatiadis (Ialysos) 88
museum visitors, numbers 25
Mycenean civilizations, existence 27–28

National History Museum of Crete 88
National Technical University of
Athens (NTUA): holistic
cultural project, parameters 45;
laboratories 40, 47, 65
NATURA 2000 network 85
Natura 2000 Network sites list 41
natural heritage site, Lavreotiki region
acceptability 85–86
Natural History Museum of the Lesvos
Petrified Forest (Lesvos) 88
natural monument 15–16
natural resources: consumption 51;
geo-diversity factors 15;
impact 4
nature conservation, integration 15
neoclassical architecture, example 60
neoclassical buildings 74
Neolithic Age 14, 76, 89
Nord-Pas de Calais-France Mineral
Basin 23

Old Coal Mine Museum 24
Old Market 74
old mine shafts 77–79
old plants, consideration 60
open-air museums, cultural heritage
tourism (relationship) 25
open pits 13, 64; concordance 22;
small-scale open pit mining/
mineral processing 80
ore deposit formation 71
ore extraction (FMC) 82
organizational diagrams, creation/
development 45

Paleolithic Age 14
Para-otokton series, carbonate formation 81

Parthenon columnar temples, expense 33
Peloponnesian War 28, 34
Pericles: Golden Age 1; monuments,
    building 29
Persian Empire (war) 29
Persian Wars, impact 30, 33
Philomouses Association, buildling
    (neoclassical architecture) 60
philosophy, evaluation 51
pilot projects, determination 22
Piraeus Maritime Tradition Museum,
    visitation 70
Plaka mining center 31
post-restoration LTCP: area A buildings
    47; outlook *57*
Prehistoric Ages 27
prehistoric times: mineral deposits,
    activity 14; mining activities
    (Lavrion) 77
production: heritage 13; intangible
    production techniques 13;
    techniques 7
Promachos, gold/ivory statue (costs) 33
Promotion of Geosite-Geoparks,
    Contribution to Sustainable
    Development 85
promotions, making 22
public institutions, sponsorship (project
    financing) 22
public-sector funding, impact 18
pyrite 3, 81, 83–84

rehabilitation: plan 65, 67; projects
    (Lavrion) 64–67
religious tourism 15
renewed buildings, impact 44
repurposing, usage 10
residential areas, creation 10–11
restoration plan 55–56
restored buildings *54*, 60, 62
reuse process, projects (conducting)
    46–51
Rhodes Aquarium (Rhodes) 88
risk management plan, design 64–65
rolling mill, lead sheet production 85
Roman Period 30
Roux-Serpieri-Fressynet 30–31, 84
Ruhr coalfield 18–19
ruins (industrial heritage concept) 8
rural Keratea, shafts/adits 78

Salina Turda Salt Mine 23
Salt Cave Museum 24

School of Architecture 38
School of Mining and Metallurgical
    Engineering 38
scientists, support 46
Second World War, mining decline 32,
    35, 37, 64
Seegrotte Underground Lake and
    Historic Gypsum Mine 23
sentimental values, tangibles values
    (relationship) 8
Serpieri, J.B. 1, 30, 31, 84; J.B. Serpieri
    Monument 74; min Shaft No. 1,
    construction 77
silver: cerussite-originating silver
    carbonate, accumulation
    81; deposits, discovery 34;
    extraction 79–80; Lavreotiki
    supplier 74; lead-originating
    silver carbonate, accumulation
    81; low-grade tailings 82; mine
    sites 2; ore, usage 1; production
    (Classical Greek Period) 82;
    supply 29
silver-containing lead (galena) ores,
    extraction 30–31
silver-lead ore, processing 77
silvery lead: ore, extraction 81;
    production 31–32
small-scale open pit mining, focus 80
smithsonite 80–81
social memory 10
socio-economic crisis 37–38
socio-economics, Lavrion mines (effects)
    33–35
soil, decontamination 46–47
Soil Rehabilitation and Regulation
    Project 66
Soil Rehabilitation and Supplemental
    Infrastructure 2
Sounion National Park 6, 41, 64, 71, 76
Soureza mining center 31, 76–77
South Africa, Big Hole (UNESCO World
    Heritage List) 23
sphalerite 3, 81
spiral-type processing plants, rarity 83
stakeholders: integration 17; project
    stakeholders, problems 47
St. Barbara's Church 60, *61*
Stone Age 27; initiation 86
stone buildings (LTCP location) 42
storage plan 67
students, temporary recruitment 70
sulfur deposits, usage 82

tailings 81–82; stacks, re-melting 1
tangible values (industrial heritage
    concept) 8
technology production, continuation 54
Tentative List of UNESCO World
    Heritage Sites 3, 6, 41–42, **42**, 90
The International Committee for the
    Conservation of Industrial
    Heritage (TICCIH) 9
Themistocles 33
thermal mineral processing 84
Thoricos Gulf rehabilitation project 65
Thorikon, ancient mine center 33
Thrace, silver/gold deposits 34
Tour-Ed Coal Mine Museum 23
tourism planning 25; geotourism
    planning 20
Town Hall 74
tunneling methods 13

underground adits, entrances 3
underground mining method, usage 69
underground mining operation 24
underground repository: construction
    67; design 67, 69; LTCP
    underground repository,
    development 67–68; waste,
    transfer 66
underground storage: construction
    works *67*; LTCP underground
    storage, development 67–68;
    in situ underground storage
    arrangement 67
underground waste repository, features
    67–69
UNESCO Global Geopark 3, 20, 22, 86
UNESCO World Heritage List 20, 23

UNESCO World Heritage Sites 9, 12,
    85–86; Lavreotiki preservation/
    inclusion, legislation 41–42;
    tentative list 41, 42; *see also*
    Tentative List of UNESCO
    World Heritage Sites
urban conservation 91; awareness/
    understanding 4, 11; locational
    studies 91
Urban Control Zone for the Lavreotiki
    Area 42
urbanization 8–9; representatives 7
urban memory, visual/spatial traces 11
urban planning, tangibles values
    (relationship) 8
uses (industrial heritage concept) 8

Venizelos, Eleftherios 60, 71
ventilation 59, 69; problems, absence 78

waste (covering), clay layers (usage) *65*
waste, components 66
water reservoirs 74
West Devon-England Mining District 23
WieliczkaSalt Mines 18, 23
workplace environment, creation 44
World Heritage Committee 9
world heritage sites 14

zinc: carbonated zinc, accumulation
    81; grade percentage 78;
    lead-zinc deposits 82; levels,
    soil analysis 67; mixed sulfide
    lead-zinc-iron-copper ores 80;
    mixture 80, 81; preparation 84
Zoological Museum of the University of
    Athens 88

For Product Safety Concerns and Information please contact our EU representative GPSR@taylorandfrancis.com
Taylor & Francis Verlag GmbH, Kaufingerstraße 24, 80331 München, Germany